Bird Census Techniques

Bird Census Techniques

Colin J. Bibby, Neil D. Burgess

Royal Society for the Protection of Birds
The Lodge
Sandy
Bedfordshire

and

David A. Hill

British Trust for Ornithology
The Nunnery
Nunnery Place
Thetford
Norfolk

Illustrated by Sandra Lambton, RSPB

Published for the
British Trust for Ornithology
and the Royal Society
for the Protection of Birds

ACADEMIC PRESS
Harcourt Brace & Company, Publishers

London · San Diego · New York
Boston · Sydney · Tokyo · Toronto

ACADEMIC PRESS LIMITED
24–28 Oval Road
London, NW1 7DX

US edition published by
ACADEMIC PRESS INC.
San Diego, CA 92101

A catalogue record for this book is available from the British Library

ISBN 0-12-095830-9

Typeset by Paston Press Ltd, Loddon, Norfolk
Printed in Great Britain by The University Press, Cambridge

Acknowledgements

The authors would like to thank in particular Dr Rob Fuller of the BTO for critically reading the whole book, Mrs Sandra Lambton of the RSPB for drawing the majority of the figures and Dr Jeremy Greenwood of the BTO for reading the proofs. The following critically read drafts of various chapters: Dr Adrian del Nevo (RSPB), Dr James Cadbury (RSPB), Mr Jeff Kirby (Wildfowl and Wetlands Trust), Mr Leslie Street (RSPB), Mr John Marchant (BTO), Mr Dave Allen (RSPB), Mr Geoff Welch (RSPB), Mr John Wilson (RSPB) and Mrs Diana Ward (RSPB). We thank them for their improving comments, however, any errors remain our own.

Mrs Anita McClune and Mrs Helen Morrow (RSPB), and Dr Rowena Langston (BTO) assisted with administrative aspects of the production. Mr Ian Dawson and Mrs Lynn Giddings (RSPB library) tracked down many references and figures. Neil Burgess and Colin Bibby acknowledge the institutional support of the RSPB and David Hill acknowledges the institutional support of the BTO.

The following also provided advice and criticism: Dr Gareth Thomas (RSPB), Dr Graham Hirons (RSPB), Dr Ceri Evans (RSPB), Dr John Cayford (RSPB), Mr Chris Mead (BTO), Dr Will Peach (BTO), Dr Steven Carter (BTO), Dr Paul Green (BTO), Dr Peter Robertson (Game Conservancy Trust), Dr Nigel Clark (BTO), Dr Steve Tapper (BTO), Dr Dick Potts (Game Conservancy Trust) and Mr Robert Petley-Jones (English Nature).

The following publishers gave permission to reproduce figures: Academic Press Ltd, American Ornithologists Union, British Trust for Ornithology, Blackwell Scientific Publications Ltd, British Birds Ltd, British Ecological Society, Cambridge University Press, Canadian Wildlife Service, Chapman and Hall, Cooper Ornithological Society, Devon Birdwatching and Preservation Society, Game Conservancy Trust, Gauthier-Villars, Harcourt Brace Jovanovich, Institute of Terrestrial Ecology, Macmillan Magazines Ltd, Ornis Scandinavica, T & AD Poyser Ltd, Royal Society for the Protection of Birds, The Wildlife Society, Wildfowl and Wetlands Trust, US Fish and Wildlife Service.

About the Authors

Colin Bibby is Director of Research at the International Council for Bird Preservation where he is interested in the role of birds as indicators for global biodiversity conservation. While Head of Conservation Science at the Royal Society for the Protection of Birds, he was a co-author of Red Data Birds in Britain. In both capacities, he has been struck by the small number of birds of conservation concern which have been counted adequately. He has counted birds in Britain and Europe, as a professional, as an amateur participant, and as an organiser of surveys for the British Trust for Ornithology. He was motivated to start this book by the belief that bird-watchers would contribute more to conservation if they put more effort into counting birds, but lack of guidance on methods was a handicap.

Neil Burgess graduated in Botany from the University of Bristol in 1983. He undertook a PhD on the evolution of the earliest land plants (400–430 million years before present) at the Natural History Museum in London, and Cardiff University in Wales from 1983 to 1987. He worked in Ecology, Research and Advisory Departments at the RSPB until 1991, producing botanical surveys of several RSPB reserves, and case studies of the methods and results of management of British habitats for birds (particularly wetland and heathland habitats). Since 1989 he has also been involved in scientific and conservation work in East Africa and since 1991 he has been employed by the RSPB International Department to manage conservation projects in Africa. This work forms part of the International Council for Bird Preservation's Africa Programme.

David Hill received his doctorate on the population ecology of wildfowl from the Edward Grey Institute of Field Ornithology at Oxford in 1982 and moved to the Game Conservancy as Head of Pheasant Research. Much of the work involved studies of marked birds, their behaviour and habitat preferences in relation to land use, largely farming, forestry and conservation. In 1987 he joined the Royal Society for the Protection of Birds as Senior Ecologist, a post which involved setting up monitoring and experiments on reserves, bringing him into close contact with sampling different bird species and habitats. In 1989 he joined the British Trust for Ornithology and established their Research, Development and Advisory Service, with responsibility for setting up and running research contracts on estuaries, farmland, woodland and uplands, using the methods outlined in this book. He is a council member of the British Ecological Society and a member of the Scientific Advisory Committee of the Wildfowl and Wetlands Trust. He has been interested in wildlife since as early as he can remember and has published scientific papers and books on the ecology of gamebirds, wildfowl, estuary waders, bird communities of native pine forest and grazed forests, insects in coppiced woodland, population dynamics, population and resource modelling, and bird distributions in relation to land cover determined by satellite imagery. Since leaving the BTO in 1992, he now runs an ecological consultancy, Ecoscope Applied Ecologists.

Contents

6. Catching and Marking

10. Description and Measurement of Bird Habitat

Preface

The idea for this book came from a general perception amongst field workers of the British Trust for Ornithology (BTO) and Royal Society for the Protection of Birds (RSPB), that the methodology for various bird census methods was scattered throughout the scientific literature and difficult to access. Indeed many field-staff or volunteers were only familiar with one or two methods and there was often little attempt to standardise methods or to check that the counting was being undertaken in a systematic manner over wide geographical areas. It was also evident that there was a need amongst volunteers, junior researchers, students and scientists in the developing world for a practical guide synthesising all aspects of the various methods of counting birds, their uses, and the things which ought to be considered before, during and after counting birds.

Methods described in the literature for counting birds are many and various. In this book we have devoted whole chapters to the most widely used and most suitable methods, and attempted to amalgamate other counting methodologies into major groups. Examples of the use of methods are provided wherever possible and the relative value of various approaches for answering specific questions is also addressed.

We have not attempted to obtain and review all the available literature on bird counting. Instead we have extracted examples in the hope of providing guidance and making workers more aware of the problems involved in bird counting.

1

Purpose and Design in Counting Birds

Introduction

Birds are counted for a wide variety of reasons by a bewildering range of methods. This guide is offered largely to people lacking the time or facilities to read and assess the extensive and often conflicting bird census literature. For a critical review of methods, the reader is referred to Verner (1985). Ralph and Scott's (1981) book contains a wide collection of papers on many aspects of the problems of counting birds, and Wiens (1989) provides abundant illustration of the extent to which methodology is critical to the valid interpretation of the results of bird counts. A pessimist reading some of this material would conclude that bird counting is so unreliable as to be of limited value. We have tried to take a more positive view. There are practical reasons for counting birds and the pitfalls in methods and study design can be avoided. It is hence possible, with adequate care and awareness of the possible limitations of the methods used, to produce valuable results.

Appropriate methods for a particular study become more obvious if there is a clear purpose, specified in advance. A recurrent theme of good study design is that methods must be tailored to aims. It is a mistake to rush to the field and start counting without prior thought. Many bird counts have no doubt been done at considerable cost of time in the field but have turned out to be just a waste of time. Results may fail to meet the aims if these are not properly understood until it comes to analysis and writing. Writing, in the sense of communicating the results to someone else, should be the final intention of any study. Because appropriate census methods are so much related to aims, it is worth giving some thought to the classes of possible aims.

Questions about numbers of birds vary greatly in scale. Are the results intended to apply to a wide geographical area or to a single site? Are many species involved or just one? Are accurate counts needed or will relative counts or presence and absence data suffice? The available effort for counting is usually limited but may be adequate for studies of a single site or species which can be more intensive than is possible if many species or large areas are involved. Accurate counts are often very difficult to obtain but for many purposes are not really needed. The key to a good study lies in recognising what kinds of data are required and understanding the pitfalls of the possible

1

counting methods. It is easy to agonise over errors of 10–20% when it may often be enough that counts are right within 100–200%. Questions about accuracy are so important that the next chapter is devoted to them. In truth, the perfect bird count probably does not exist, but this need not prevent the extraction of useful results from a good study. For a review of data analysis and the design of experiments in ornithology the reader is referred to James and McCulloch (1985) and Hairston (1989).

Stock-taking

The simplest aims are of a kind that ask what birds occur here. Such studies might be used for evaluating poorly known sites or for setting a baseline prior to fuller study or experiment. An answer that is little more than a species list, perhaps with gross approximations of numbers between common and rare, may be sufficient. Elaborate methods may not be necessary and it is surprising how informative such studies can be. On a wide scale, such are the aims and methods of atlas studies, but similar work is possible at a more local level with a finer grid (Chapter 9). The general presumption of such an approach is that presence or absence can be detected with some reliability given enough effort. Consideration might be given to problems of counting particular bird species (Chapter 7) or dealing with large numbers (Chapter 8). It will almost never be possible to do equally well for all species. Indeed, such an aim is almost hopeless; some birds are very difficult to detect, let alone count.

Stock-taking aims are very often appropriate for studying single species on an extensive scale. The methods may be fairly crude but nonetheless valuable, especially if it is reasonable to expect to count the whole population of a wide area. Many surveys of single species requiring help from a large number of people have taken such an approach. It is very difficult to know how accurate the results are because both gaps in the coverage and flaws in the methods may be poorly known. A common assumption is that features of interest will be clear enough to survive some deficiencies of method. Thus population levels of Herons (Reynolds 1979) and Rooks (Sage and Vernon 1978) have been monitored by fairly simple counts and with useful, even if not totally accurate, results. Waders which are difficult to count have been studied both in breeding (Smith 1983) and non-breeding (Prater 1981) habitats. The ideal in such studies is to have a uniform effort in different areas. Failing this, the effort needs to be known, especially if some places are not visited at all (Box 1.1).

Population counts with varying effort.

Box 1.1

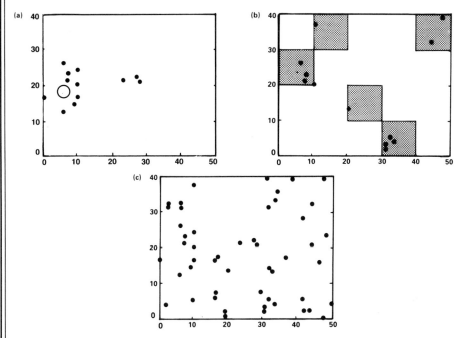

A hypothetical county happens to be a perfect rectangle and the recorder is interested in describing the Magpie population. Three different approaches are tried in succession.

(a) *Map of all Magpie nest areas in the past 5 years*: what does it show? A possibility is that 10 km squares 01 and 02 are good for Magpies which are generally rare in the county. Perhaps there are about 20 pairs in the whole county allowing for a few which have been missed?

(b) *Sample survey*: five randomly selected 10 km squares were fully covered and the 12 nests marked were found. What does this mean? Square 02 now looks less special. Four nests were also found in square 30 where Magpies had not been recorded casually as shown above. The total population for the county can be estimated from this survey as $12 \times 20/5 = 48$ pairs. These squares could be re-visited in future years to assess population changes.

(c) *A full survey*: this shows that there were actually 50 pairs. Assuming that the method was accurate, this figure is correct.

The sample survey took only a quarter of the effort and produced quite a good answer for the total population but did not, of course, actually locate most of the pairs. The sample survey was good enough to get total numbers but not sufficient if location was also important.

The data from the full survey could be analysed to explain variation of density in terms of habitat.

Example (a) was totally wrong. This was because the ringing group based in the town marked with a circle was very keen on Magpies and submitted all the records. No one else thought Magpies worthy of note!

**Box
1.2**

Mapping census and habitat map.

(a)

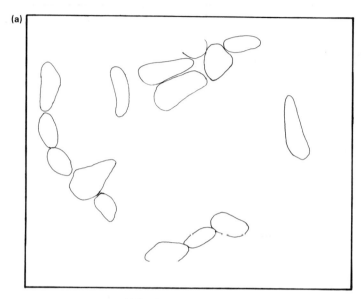

Yellowhammer territories

(b)

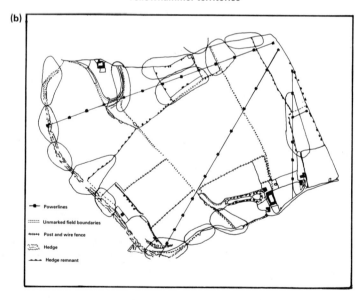

- ● Powerlines
- Unmarked field boundaries
- Post and wire fence
- Hedge
- Hedge remnant

(a) Map of Yellowhammer territories on a mapped plot (Williamson 1968). It means very little on its own if you happen not to know the site where it was done.
(b) It makes more sense when some key habitat features such as field boundaries have also been mapped in. Yellowhammer seem to prefer territories along hedges or other field boundaries, with few in the centres of fields.

Mapping distributions at different levels of scale.

Box 1.3

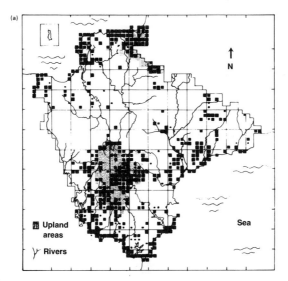

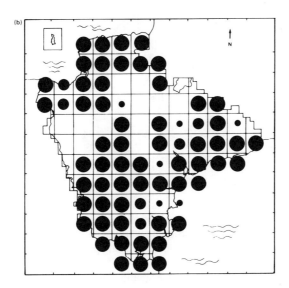

(a) Distribution of Stonechat from the Tetrad Atlas of Breeding Birds in Devon (from Sitters 1988). Stonechats occur especially on the upland areas of Exmoor, Dartmoor and in coastal regions. The survey results are presented in 2 × 2 km squares. A repeat survey will be able to show any distribution changes.

(b) The same data redrawn by 10 × 10-km squares which is the standard used for national atlases in Britain (e.g. Lack 1986). The details of factors influencing distribution have been obscured. It will take a much bigger range change before this scale of work can detect it.

Box 1.4

Survey coverage affecting results.

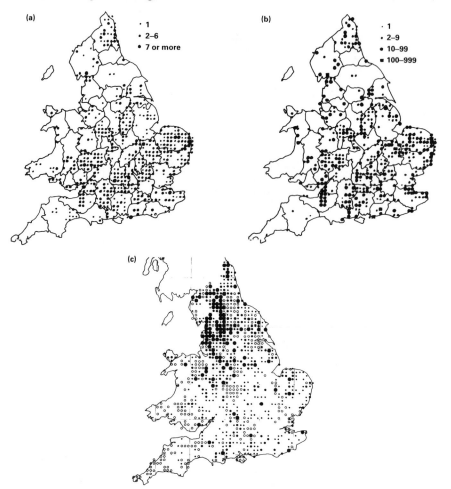

(a) The distribution and numbers of grassland sites surveyed in each 10 km square. All survey sites are included even if no breeding waders were located (from Smith 1983).
(b) The distribution and numbers of breeding Lapwing pairs found on grassland in each 10 km square (from Smith 1983).
(c) The distribution of nesting Lapwings in England and Wales in 1987 (from Shrubb and Lack 1991). The symbols represent numbers in one tetrad in each 10 km square: small dot = 1–4 pairs recorded; medium dot = 5–10 pairs; large dot more than 10 pairs; blank = the tetrad was visited but held no birds; and an open circle = not visited.

What does it mean? It is very difficult to infer much from the bird map (b) because it is influenced by the coverage map (a). The coverage map may represent the distribution of suitable habitats but this is not claimed because coverage was incomplete. The species distribution map (b) does not, in fact, represent the breeding distribution of the Lapwing which reaches its greatest abundance in the north-west in England (c).

Box 1.5

A complete population survey.

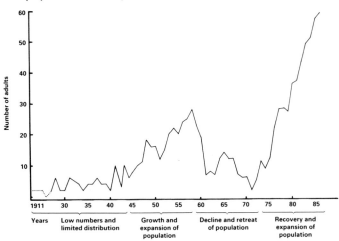

The numbers of breeding adult Marsh Harriers in Britain, 1911–1986 (from Day 1988). Problems with sample coverage and design do not arise in interpreting what this means. The only question is whether the coverage was complete for a study dating back so far. It is believed that it was for such a conspicuous bird which breeds in a restricted range of habitats.

Box 1.6

Trends in population derived from indices.

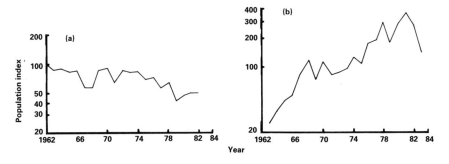

Trends are shown for breeding Lapwing populations in areas of England and Wales dominated by cereals (a) and sheep rearing (b) (from O'Connor and Shrubb 1986). The trends are derived by estimating the proportional population changes from one year to the next on plots that were counted by the same observer and methods in both years.

The implication that these are general results seems surprising from casual observation, which suggests that Lapwings have disappeared very widely from cereal areas but have also declined markedly in sheep-dominated areas such as Wales. The generality of the results depends on how the study plots were chosen and where they were located. This information was not given. Without it, it is not possible to say exactly what these plots mean.

Box 1.7

Population trends from studies using paired plots.

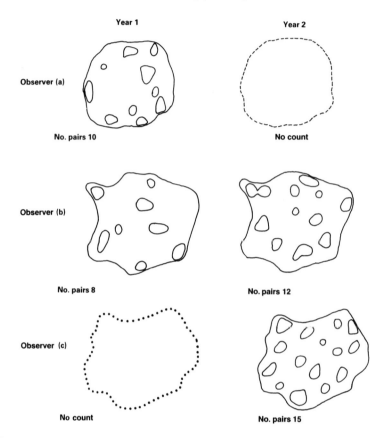

Three mapping plots (a–c) in one county are studied in 2 years. All plots are 25 ha in extent. Hypothetical results show territories of Wood Warblers. What does it mean?

The mean density of Wood Warblers was 36 pairs per km² in year 1 and 54 pairs per km² in year 2. There was a 50% increase from year 1 to year 2 as shown by mean density on all the plots studied in each year. As it happens, the one plot studied in both years also showed a 50% increase. So might we infer that Wood Warblers increased in numbers by 50% in the county from year 1 to year 2 and that they generally occur at densities of 30–60 pairs per km²?

Not necessarily. The observers chose their study areas and picked interesting-looking woods, not typical of the county where few of the woods actually support Wood Warblers at all. So no reliable general estimate of Wood Warbler density in the county's woods can be made. Half of the woods in the county blew down in a gale in winter of years 1 and 2, which is why observer (a) gave up his plot in year 2. The total number of Wood Warblers in the county actually went down by 25% between the two years. This is not represented in

Box
1.7
cont.

paired plot studies of this kind if observers stop counting on plots that have become unsuitable.

This illustration is deliberately exaggerated to make its points. It would be obvious that half the woods had blown down. It is easy to see that more subtle changes could be overlooked and the results could mislead.

Box
1.8

A hypothetical study designed to show the effect of age of trees on bird communities in conifer forests produced the following results from four equal-sized mapped plots.

Plot	A	B	C	D
Age (years)	2	5	10	40
Tree Pipit	0	6	1	18
Nightjar	0	0	3	0
Great Spotted Woodpecker	1	0	0	0

It is impossible to tell what the data mean! The plots differed in many attributes, other than age, which might have caused the differences. Plot D was surrounded by a large clear-fell which attracted many Tree Pipits to its edge. Plot C was the only one on a sandy site which is why it had Nightjars, the rest were on clay. Plot A was adjacent to an oakwood which is where the Great Spotted Woodpecker nested. The possibilities for explaining the results are endless. The inference that Tree Pipits favour older stands and Great Spotted Woodpeckers prefer very young trees would be wrong.

What could be done to improve the study?

(1) Match the plots more carefully to reduce the number of factors in which they differ.

(2) Spread the survey more widely so there can be replicates for areas of different age. Using point counts or transects would allow this.

(3) Study a small number of plots for several years, though 40 years would be a long time to wait for the results.

Box 1.9

The use of an experimental control.

Plot A

Year A
Before thinning

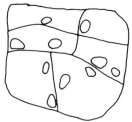

No. pairs 10

Year B
After thinning

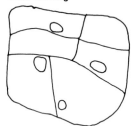

No. pairs 4

Control plot

Year A

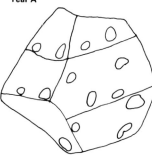

No. pairs 14

Year B

No. pairs 15

The figure shows a hypothetical experiment to test the effects of thinning of conifer forest plantations on populations of Wrens. One plot (A) was studied in the years before and after thinning.

What does the decline in numbers mean? It is not clear. The snowfall in December was the heaviest in living memory and Wren numbers may have suffered even if the forest had not been thinned. But the national Common Birds Census index went up by 8% from year A to year B. Unfortunately this is not very helpful as the study area is in northern Scotland and the CBC index is biased to southern England which did not have much snow.

What could be done to improve the study? Include a control plot without any vegetation change. This will measure the effects of other factors such as weather. In the control plot in this study, the numbers of Wrens did not decline.

Conclusion: Wrens did not suffer severe mortality in the great blizzard (it thawed very fast). The decline in numbers was therefore probably due to the thinning, but without replication it is impossible to say how repeatable this result might be.

If the area of study is relatively small and special such as a nature reserve or a particular wood, farm or marsh, then mapping methods (Chapter 3) may be used to answer stock-taking questions about the numbers and distribution of breeding birds. The relatively high costs of time per unit of result may not matter. The great advantage of mapping is that the result is not only a table of numbers but also a picture of the approximate location of the bird territories. Someone who knows the site will find the species maps of considerable interest. An outsider will find them hard to interpret unless quite a good habitat map is also made (Box 1.2). A common use of such maps is to estimate what losses of habitat or drastic habitat change might produce an effect on birds. One can show which parts of a plot hold particular birds and then argue that the loss of these areas would have a big impact on the local birds. If a bigger area or a result of wider generalisation is needed, then transect or point counts might be appropriate (Chapters 4 and 5).

Distribution studies

Several aims do not require counts at all but merely to specify where birds do and do not occur. Distribution maps used to be drawn from casual records. More recently, they have mapped presence or absence by some map unit: 10 km squares in Britain nationally and 2 km squares for various counties. To be interesting, such maps need to have a mixture of areas with and without birds, which depends to some extent on the distribution of the bird and the size of the map units (Box 1.3). Coverage needs either to be reasonably uniform or to be measured and reported (Box 1.4). Otherwise, the resulting maps will show distribution of observer effort as much as distribution of birds. Distribution studies are described in Chapter 9. Various kinds of counts might be based on separate and recognised areas so that results can be expressed as maps of relative density as well as tabulation of numbers.

Population monitoring

Trends in numbers over time are of particular interest to nature conservation. The birds may be inherently rare and thus in need of surveillance. They may occupy habitats known to be changing, or perhaps be candidates for recognising adverse effects of pesticides or pollution. Numbers also fluctuate naturally, usually because of the effects of weather on reproduction and survival, but also because of the density-dependent effects of population level itself. It is necessary to understand such fluctuations so that they are not confused with those attributable to human beings. The ability to distinguish between natural and man-made population changes is an essential attribute of a successful monitoring scheme (Baillie 1990). The expectation in such studies is that annual changes in a long-term trend may be quite small but there may be substantial year-to-year variation which could conceal a long-term trend. The fluctuations due to weather or population level need to be measured with some confidence if they are to be recognised.

Species in which the whole population can be located with reasonable confidence are not much of a problem. Thus the number of Marsh Harriers breeding in Britain has been fairly well documented over a great many years (Box 1.5) as a result of many people checking the suitable patches of habitat within the known range each year. Almost complete counts have been made at periodic intervals of birds as diverse as Great-crested Grebes, Peregrines, Gannets and Dartford Warblers. In all these examples, valuable results have been obtained in spite of methodological imperfections.

If the species is more numerous or widespread, population monitoring may be no less important but complete counting may be out of the question. Moreover, these species often give greater problems in counting. It is relatively easy to tell if a crag has a pair of Peregrines or not. It is substantially harder to tell if a wood has ten pairs of Chaffinches or 20 pairs. Such problems of population monitoring fall into two categories. First, if a complete count is out of the question, a sample is required and this must be selected with some care. Random sampling of 10 km squares (or similar units) is possible, especially if prior knowledge of which are or are not occupied is available. Such a system has been used with Mute Swans (Ogilvie 1986) and Wood Warblers (Bibby 1989). Samples of auk breeding colonies (Stowe 1982), on the other hand, have been criticised because their distribution was not random. An important point is that, if sampling is non-random, results can less justifiably be generalised to infer what the population as a whole was doing. It is thus important to understand just what has been sampled and how it might relate to the whole population in relation to such factors as geographical spread and habitat coverage.

The most extensive sample population monitoring scheme in Britain is the Common Birds Census (CBC). This illustrates this sampling problem well because it is tempting to believe that the published indices for farmland and woodland apply to the whole populations (Box 1.6). In fact they apply very strictly to the places from which they came, which are strongly biased to south and east England (Fuller *et al.* 1985). Internal checks are made to ensure that, within this area, the plots are representative of the range of agricultural practice and do not drift over time. One could imagine observers giving up when their farm is converted into a wheat prairie, while a new starter might be tempted to select somewhere more well-hedged. It is easy to see how a non-random sample could fail to show up the effects of habitat deterioration for such a reason (Box 1.7). In practice, random sampling is often not possible, so one must be well aware of the potential consequences of this departure from the ideal.

The second problem in population monitoring is that the methods should be repeatable from year to year. Repeatable counts do not need to be accurate in the sense that the population numbers recorded are the actual population figures. If the numbers are constantly wrong for any reason, the changes from year to year can still be measured accurately. The CBC tidily avoids this problem by calculating a population index based only on plots counted in the same way by the same observer in successive years. Provided an observer with poor hearing detects twice as many Goldcrests if numbers double, it does not matter if only two and four pairs are recorded when in fact there were ten and 20. If, on the other hand, the counter becomes gradually deafer over time, Goldcrests might appear to become rarer when in fact they have actually maintained their numbers.

For songbirds in Britain, there has been a general presumption in favour of mapping methods because of the example of the CBC. Other methods such as transects (Chapter 4), point counts (Chapter 5) and capture methods (Chapter 6) are all routinely used elsewhere. Even the remarkably ill-standardised Christmas Bird Count in the USA has been able to detect population trends (Bock and Root 1981; Drennan 1981; Root 1988) although this was not its original purpose. In this case, confidence in the results is clearly less than with other techniques, so only larger changes can reliably be believed to be real. The most important point about population monitoring is not that the methods be accurate but that they be similar from year to year so that systematic inaccuracy does not have an adverse effect. The other essential consideration is that the areas being counted are representative of the population to which the answer will be generalised. Such studies need to have a wide spread of samples to be meaningful.

Assessment of habitat requirements

The habitats of birds are of interest for a variety of reasons. In the applied field, predicting the effects of land-use changes is often important. Results may also have application in land-use planning issues or in the management of areas such as nature reserves specifically for birds. Legislation requiring developers to make environmental impact assessments of their plans has resulted in considerable work on bird habitats in the USA from which much useful methodology has arisen.

Greater accuracy may be required to permit comparison with results previously published (though there is rarely much certainty that these will be accurate anyway). Comparison internal to the study may also pose problems of accuracy, especially if the range of habitats covered is large: e.g. are counts in dense habitats as good as those in open ones or are differences partly an artefact of census problems? Not only may the requirements for accuracy be greater but so too may be the demand for quantity of data and these may tend to suggest divergent options in the study design.

The reason for needing copious data is that a small number of plots will vary from one another in a great many attributes. Some of these could produce spurious or confounding results. A small set of woods might, for instance, be chosen to represent a range of ages. But what if one happens to face south, or be on better soil, or have less scrub, or a more open canopy? Clearly in a small number of plots, the prospect for any trend under investigation to be swamped by other variables is great. If the other variables were not recorded at all, this could go unnoticed (Box 1.8). If they were recorded, then the statistics will not work unless there are many more plots than habitat variables (see Chapter 10). The other reason why data are required in quantity is that, in a small plot, the majority of birds will belong to a small number of common species and there will be rather few records of a greater number of species which are each scarce. It may, however, be the scarcer species that are more exacting in their habitat requirements and of more interest to the study. More efficient methods such as point or transect counts may be preferred to tackle this problem. A special modification of the point counting system can be used to study bird habitats without any counting of the normal kind at all (see Chapter 5). Habitats may be studied through a mapping census by looking at the distribution of birds within a plot in relation to vegetation features (see Chapter 3).

Habitat-oriented studies also pose problems in recording the habitat. They clearly cannot work adequately without proper recording. This problem is discussed in Chapter 10. Even after the habitat data have been collected in a suitably designed study, there are difficult analytical problems. Computers and elaborate multivariate statistical methods may be required. This area is largely beyond the scope of the present book but a brief discussion is provided in Chapter 10.

Management experiments

An experimental approach to test hypotheses about bird habitats has a lot to be said for it, apart from the problem of manipulating vegetation on a sufficient scale. Observation trials are more often conducted by taking advantage of some planned change, be it deliberately intended to encourage wildlife on a nature reserve or possibly prejudicial as in a so-called improvement scheme on a river. To be fully satisfactory, it is essential to know what would have changed on the plot if the alteration had not occurred (Box 1.9). The before and after measurements may be made in successive years. But how can you tell whether a change in a particular species was due to the experimental factor or to an extraneous factor such as poor rainfall in the winter quarters?

A well designed study needs its own control with the same level of census effort in the before and after stages of the trial as is given to the study plot itself. A particular problem occurs in studies conducted in a single season, an approach often necessary in the investigation of pesticide applications. It then becomes very difficult to deal with the seasonal changes in detectability of birds. Only very detailed methods can adequately approach this kind of problem.

A frequent difficulty with studies of this kind is that samples are too small both in the statistical sense and by way of only treating the more common species. If the study is being done by expedience with the management already planned, it is worth considering whether the area is big enough. In coppice areas in woods, for example, the plots cut are rarely larger than a hectare, so most possible species are absent or represented by a single pair whose territory boundaries may well be wider than the plot. If the area is too small to show any meaningful results, the study needs redesigning before any data are collected.

Conclusions

Throughout the rest of the book, we have attempted to illustrate the use of particular methods to meet particular aims. Our examples are necessarily very selective. Table 1 indicates a range of recent studies classified according to their methods and purposes. This shows where to look in the book for further discussion and also offers some pointers for additional reading.

In a brief sketch, this chapter has given a range of possible objectives of a census study. The rest of the book describes a wide range of methods. There is a challenge in picking the right method. It will be a recurring theme that the right method is one that is appropriate to the questions being asked. Care in thinking about matching methods to aims will more than repay the time that could otherwise be wasted. So the first task is to specify just what the question is. What scale of result is needed? Must it apply to one area, to one habitat type or to many habitats and a large area? The selection and number of study plots will largely define the scale of the work but do they match the question? Is a complete survey needed or is the proposed scale so large that sample plots will have to be used? If samples are to be used, how will they be selected to be representative of the conditions in question?

What accuracy and precision of results are needed (see Chapter 2) and how will they be achieved? An extreme view of accuracy concerns the need for absolute or relative measures. Is it necessary to know exactly how numerous a particular bird is or will an index suffice provided the index is large where the bird is abundant and vice versa? It is common to believe that absolute numbers are needed. In practice, they are at best expensive to acquire. At worst, they may be virtually impossible to achieve and so represent a futile quest. A great many questions can be answered with relative counts.

Finally, thought should be given to the methods of analysis. This gives a further chance to recognise whether the methods are likely to suit the aims of the study. Once these hurdles are crossed, the time has come to select a census method.

Table 1.1 Examples of studies using different methods to address a variety of study areas. Numbers in parentheses are the Chapter in the book where the example occurs and/or where the relevant reference for the study is given. It can be seen that some counting methods are more regularly used for addressing certain questions than others.

Counting method	Chapter	Biogeography	Studies	
			Site evaluation/ inventory	Index of population changes
Absolute	1 2		(1) Marsh Harriers in UK (Day 1988) (7) Herons (Marquiss 1989)	(1) Marsh Harriers in UK (Day 1988) (7) Herons (Marquiss 1989)
Mapping	3		(7) Bittern (7) Red Grouse (7) Nightjar (7) Owls Passerines	(1) Lapwing population in UK (Shrubb and Lack 1991) (3) CBC programme in UK (7) Little Grebe in UK (Vinicombe 1982) (7) Bittern in UK (7) Red Grouse (Hudson and Rands 1988) (7) Pheasant (Hill and Robertson 1988) (7) Owls and Nightjars (7) Coots/Moorhens (7) Corncrake (Stowe and Hudson 1988)
Transects	4	(4) Seabirds at sea (Tasker *et al.* 1984)	(7) Capercaillie (2) Black Grouse (7) Upland waders Passerines	(4) Finland Bird Census (7) Raptors in USA (Fuller and Mosher 1981) (7) Capercaillie (Rolstad and Wegge 1987) (7) Owls (Fuller and Mosher 1981) (8) Skuas (Furness 1982) (7) Upland waders (Fuller *et al.* 1983) (8) Gulls Lovebirds (Thompson 1989)

	Studies		
Habitat preferences	Density dependence	Lifespan and survival	Energetics
	Avocets (Hill 1988) Sparrowhawk (Newton 1988)		
(1) Lapwing population of UK (Shrubb and Lack 1991) (3, 10) UK woodland habitats (Fuller *et al.* 1989) (10) Grassland birds in USA (Wiens 1969, 1973)	Willow Grouse (Andreev 1988) Nuthatch (Matthysen 1989)		
(4) Shrub-steppe birds of USA (Wiens and Rotenberry 1985) Aerial seabirds (Ryan and Cooper 1989)			

(Continued)

Table 1.1 continued

Counting method	Chapter	Biogeography	Studies Site evaluation/ inventory	Index of population changes
Point counts	5	(5) Azores Bullfinch (Bibby and Charlton 1991) (Massa and Fedrigo 1989)	Passerines	(5) Breeding birds in USA (Robbins *et al.* 1986) Oak–pine woodland birds (Verner and Milne 1989)
Capture–recapture	6			(6) Lincoln index Canada Goose (Hestbeck and Malecki 1989) (6) du Feu method (du Feu *et al.* 1983) Lovebirds (Thompson 1989)
Catch per unit effort	6			(6) UK Constant Effort Sites (6) Stock Dove in UK (O'Connor and Mead 1984)
Radio-tracking	6 9			

		Studies	
Habitat preferences	Density dependence	Lifespan and survival	Energetics
(5) Young forestry plantation in UK (Bibby *et al.* 1985) (10) Woodland birds and structure (Bibby and Robins 1985; Hill *et al.* 1990)			
(6) Pheasants in UK (Hill and Robertson 1988)		(6) Bewick's Swans at Slimbridge, Barnacle Goose (Owen and Black 1989) White Stork (Kanyamibwa *et al.* 1990) MULT program (Conroy *et al.* 1989)	
(9) Pheasants and woodland edge (Hill and Robertson 1988) (10) Grey and Red-legged Partridge (Green 1984) (10) Woodcock (Hirons and Johnson 1987)			Wandering Albatross (Jouventin and Weimerskirch 1990)

(Continued)

Table 1.1 continued

Counting method	Chapter	Studies		
		Biogeography	Site evaluation/ inventory	Index of population changes
Direct and indirect counting	7 8	Canvasback ducks (Lovvorn 1989) Penguin rookeries (Schwaller *et al.* 1989)	(7) Wildfowl, waders, corvids, seabirds (terns, auks, gulls) (10) Satellites and waders (Avery and Haines-Young 1990)	(7) Nesting female ducks (Hill 1984 a,b) (7) Off-duty male ducks (Pöysä 1984) (7) Migrating raptors (7) Grey Partridge (Potts 1986) (7) Lowland waders (BTO 1989) (7) Corvids (8) Cliff-nesting seabirds (8) Wader roosts Emu (Horne and Short 1988) Woodcock, singing ground counts (Tappe *et al.* 1989)
Look–see	7	Golden Eagle (Watson *et al.* 1989)	(7) Roosting Hen Harriers (Clarke and Watson 1990)	(7) Divers in UK (Campbell and Talbot 1987) (7) Herons in UK (Marquiss 1989) (7) Obvious wildfowl (Ogilvie 1986) (7) Buzzards in UK (Taylor *et al.* 1988) (7) Roosting raptors (Clarke and Watson 1990) (7) Owls (7) Corvids

	Studies		
Habitat preferences	Density dependence	Lifespan and survival	Energetics
Diving ducks (Bergan and Smith 1989) Black-tailed Godwit (Buker and Groen 1989)	Waders (Goss-Custard and Durell 1990) Magpie Goose (Bayliss 1989) Snow Goose (Cooch *et al.* 1989)		Diving ducks (Bergan *et al.* 1989) Waders (Young 1989)

(*Continued*)

Table 1.1 continued

Counting method	Chapter	Biogeography	Studies Site evaluation/ inventory	Studies Index of population changes
Distribution presence/ absence	1 2	(1) Lapwings in UK (Shrubb and Lack 1991) (1) Waders (Smith 1983) (1) Stonechat in Devon (Sitters 1988) (7) Buzzards in UK (Taylor *et al.* 1988)	(9) Fuerteventura Stonechat (Bibby and Hill 1987)	(9) Red-backed Shrike (Bibby 1973) (9) Wintering Birds in USA (CBC) (Root 1988) Changes in avian density (Bart and Klosiewski 1989)
	9	(9) UK Atlas (Sharrock 1976) (9) USA Christmas Bird Count (Root 1988) Birds and afforestation— satellite imagery (Smith 1988) Canada Atlas (Welsh 1989) USA—Breeding Atlas (Robbins *et al.* 1986) East African Atlas (Pomeroy 1989)		

Summary and points to consider

1. What is the purpose of the study?

Is the question worthwhile?
What can be learned from reading about previous studies?
What scale of generality is wanted?
Which and how many species need to be included?
Are guesses, indices or absolute numbers needed?
What sort of sample sizes are needed?

2. What are the field methods?

What basic methods (maps, point counts, look-see, etc.) should be used?
How many plots/routes/points should be used?
How are sample areas to be chosen?
How much will it cost (time or money)?

		Studies	
Habitat preferences	Density dependence	Lifespan and survival	Energetics
(1) Lapwings in UK (Smith 1983; Shrubb and Lack 1991)			
(1) Yellowhammer on farmland (Williamson 1968)			
(8) Waders on estuaries			
(9) Winter birds in USA (Root 1988)			
(9) Woodland birds (Fuller *et al.* 1989)			
(9) Heathland birds (Bibby and Tubbs 1975)			
(10) Partridges on farmland (Green 1984)			

Are the observers skilled or how will they be trained?
What are the likely sources of bias?
What steps will be taken to deal with bias?
How will the results be recorded—design of forms, etc.?

3. Do the methods suit the purpose?

Are the methods sufficient but not excessive?
Do any other variables need to be measured?

4. How will the analysis work?

Are the sample sizes going to be sufficient?
What about the scarcer species?
Are there enough data points to deal with many habitat variables?
What sort of statistical and computing facilities will be needed?

2

Census Errors

Introduction

The numbers of birds in a particular place, the average density of birds in a habitat for one season, or whatever is to be estimated, has a precise value — the true value, which is of course unknown to us. Except in trivial cases the estimate of this value will be somewhat different from the true value. This difference is called error, the word being used in a statistical sense rather than in its common meaning of a mistake. Many reports of bird censuses do not acknowledge errors directly but understanding them is of great importance in designing a study. Casually one might think that they are undesirable and should be avoided. In fact this is neither possible nor necessarily even the best approach. Whatever else, they should not be ignored. There are two sources of error: normal variation and bias. Results with minimal amounts of each are known as precise and accurate (or unbiased), respectively. The concepts of precision and accuracy are very important and must always be considered in designing a bird-counting study.

Precision

Imagine that we want to measure the average density of Skylarks in a relatively uniform area of arable farmland in a particular summer. Assume, improbably, that we have some means of telling exactly how many pairs there are in a small plot. The density on this plot might be quite similar to the average density but it is unlikely that it will be exactly right. In the same way, it is most unlikely that you, the reader, are of exactly average height. If we take several small plots and average the densities, the result will probably be better. If, and this is a theoretical rather than practical suggestion, we took 100 plots, the average would probably be rather close to the true density. A histogram of the frequency of small classes of densities in these plots would have an approximately symmetrical humped shape. With an infinite number of plots, this could be represented by a smooth curve: the Normal distribution. A single sample is very unlikely to be exactly representative. As more samples are added so their average becomes closer to the true value. The result becomes more precise (Box 2.1).

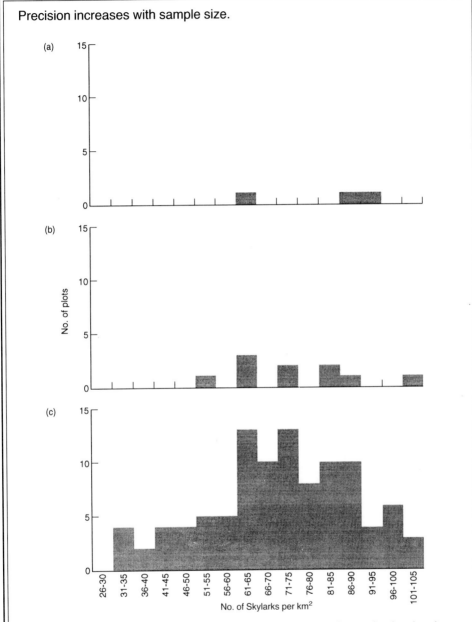

Box 2.1

Precision increases with sample size.

(a)

(b)

No. of plots

(c)

No. of Skylarks per km²

The figure shows hypothetical densities of Skylarks from small sample plots in a large uniform area of habitat. The true density is 70 birds per km², SE = 20.
(a) Three study plots show results scattered about the truth (which is unknown to the observer; sample mean 81.6 birds per km²).
(b) The mean of 10 results is quite close to the truth (sample mean = 75.0 birds per km²).
(c) One hundred plots would take a huge effort to cover but give a good impression of the mean density and its variation from plot to plot (sample mean = 71.2 birds per km²).

Box 2.2

Precision increases with sample size.

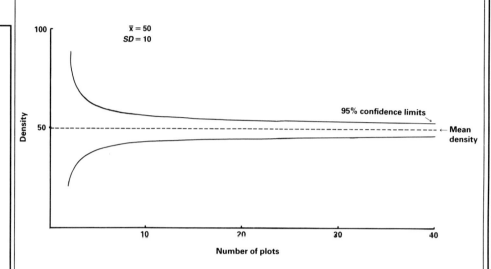

The figure shows the relationship between the number of plots sampled and the confidence which can be put on the result for a particular hypothetical population.

Box 2.3

Random and Stratified plots.

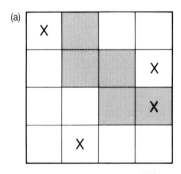
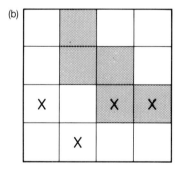

(a) The area to be studied is divided into square blocks on a map. Plots are selected at random to ensure that the results can be taken to be generally representative of the whole area under study. By chance, three plots fall in one habitat and only one in the other.

(b) The whole area is again divided into square blocks on a map. These can be attributed to each of two habitats which are known or thought to differ in their suitability for the study bird. Two randomly chosen plots are picked within each of the two categories of habitat.

A biased sampling method.

Box 2.4

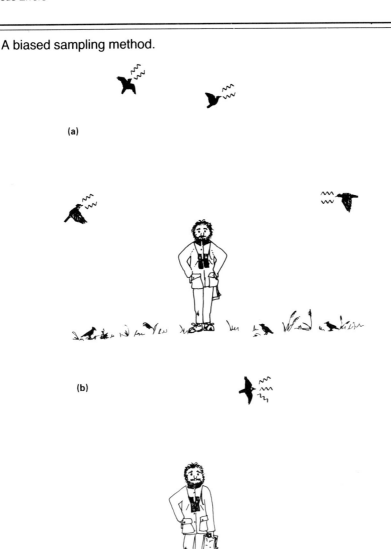

(a)

(b)

The relative density of Skylarks is measured at a sample plot by counting all the birds singing during a 10-minute observation period.
(a) A count of all the birds in the sky showed four birds, even though there were actually seven present (three on the ground).
(b) A count of singing birds showed one bird and again there were really seven present. In both cases the number of birds recorded is less than the true number. Census results will almost always be less than the real number of birds present. The result may still, however, be proportional to the true number: if there are twice as many birds in a plot then twice as many might be expected to be counted.

Box 2.5

Precision and bias.

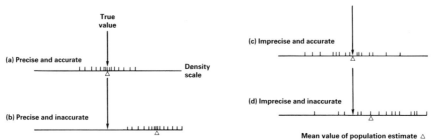

Precision and bias vary independently so either can be high or low in a particular study. The results for 15 different plots are shown in relation to the unknown true density.

(a) Precise and accurate. The results are closely spaced about the true value. This is the ideal situation and is probably rarely met in practice.

(b) Precise and inaccurate. The results are closely spaced but their average deviates from the true value. Since the true value is unknown to the observer, this result cannot readily be recognised as different from that in (a).

(c) Imprecise and accurate. The results are spread rather widely about the true value.

(d) Imprecise and inaccurate. The results are spread widely and their average deviates from the true value. Again, since the true value is unknown to the observer, this result cannot readily be recognised as different from that in (c). Some cynics would say that this pattern is the most common result obtained in bird counts.

Box 2.6

Bias due to effort and speed.

You see more birds if you put in more effort. If you race through a plot, you will miss the quiet and skulking species.

Bias due to habitat.

Box
2.7

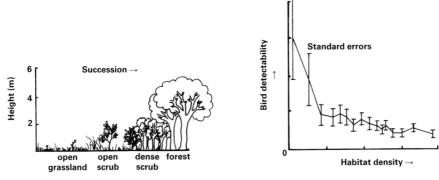

Birds are more conspicuous in open habitats than in dense woodland (from Bibby and Buckland 1987). The hypothetical species is equally abundant across the succession, but might appear more abundant in the grassland and young trees where it is more easily detected. This effect is particularly serious if the bias arises from the same source as the object of study (such as the effect of forest succession on bird communities).

Bias due to bird species.

Box
2.8

Noisy and active birds are easier to find than quiet or skulking ones. As a result, different species may in practice be counted on different scales that do not allow comparison with each other.

Box 2.9	Bias due to bird density.

At high bird densities the observer may be swamped by the numbers of birds to be located, recognised and counted. It may be difficult to separate the individuals previously recorded.

Box 2.10	Bias due to season.

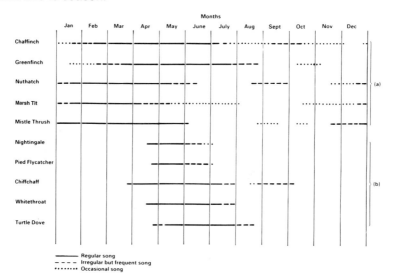

Birds sing over different periods of the year (from Alexander 1935).
(a) In the UK the resident species sing early in the spring.
(b) The resident species have all but finished singing before African migrant species arrive and begin singing.

Bias due to time of day.   **Box**
                                                              2.11

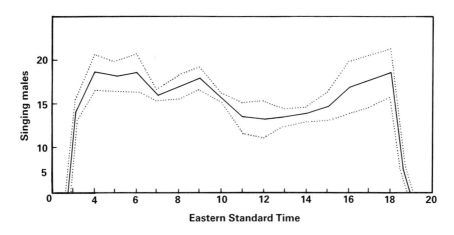

All-day activity patterns from point counts on five mid-July days. Solid line is singing males per 20 minutes, with 95% confidence limits as dotted lines.

Activity and song output is often greatest near dawn, low during the middle of the day, and higher close to dusk (from Robbins 1981). Places visited in the middle of the day will therefore appear to be poorer for birds.

Bias due to weather.   **Box**
                                                              2.12

In wet or windy weather, birds may be less active and skulk out of sight. Calls are harder to hear against the noise of the wind or rain. The observer finds it difficult to concentrate on keeping warm and dry as well as counting birds.

In this example, the error causing imprecision arose from the fact that we surveyed only a small part of the total area in question and the birds were not perfectly uniformly distributed. This might have been by chance, history of their settlement or because the habitat was not as perfectly uniform as tacitly assumed. Some kinds of measuring errors also have an identical property of averaging out in large samples. If 100 people used the same ruler to measure your height, there would be some variation in the results but they would probably be closely grouped about the true value. The Normal distribution is of such widespread occurrence as to be the basis of a large part of statistics and is dealt with in the early chapters of a great many books. These should be consulted for more than the rather rapid treatment given here. Fowler and Cohen (1986) provide an introduction with examples drawn from the bird world.

If the variation in density from one small plot to another is little to do with the census method, then it is a property of the circumstances being studied. Nothing can be done to alter this fact apart from making the plots bigger, which is equivalent to taking more samples and averaging them. The correct response is not to put all the effort into one huge plot and get a single answer but to count several smaller plots. In this way, the mean can be obtained together with an estimate of its precision or error. The standard error of densities is the measure of the error inherent in a single observation. A simple statistical manipulation with an appropriate value looked up in a table of the *t*-distribution gives an estimate of the range within which we can be reasonably confident that the true answer lies. A common conventional value for reasonable certainty is 95%. The upper and lower points of this range, within which the true value is expected to lie 95 times out of 100, are known as the 95% confidence limits of the estimate. As more samples are added, so the range between the confidence limits will tend to contract. This formally measures the intuitively obvious concept that the larger the sample size, the closer the average will be to the true value (Box 2.2).

Precision can therefore be measured. It can also be increased by taking more samples, but this takes more time. Unfortunately, precision increases only in proportion to the square root of sample size. To double the precision obtained from ten samples requires another 30. To double it again would require a further 120. Clearly this scale of multiplication rapidly becomes unrealistic. The precision that can be obtained has to be balanced against the time available. It also has to be considered in relation to the nature of the question being asked. Is an estimated density that is right to within 50% good enough? Or must it be better? If it is intended to compare the result, say with the density in a different place, then the precision required depends on how large a difference is expected. If the two densities are grossly different, then fairly imprecise estimates will confirm the fact. If they are only rather slightly different, then a lot of work will be required to get sufficient precision. In the best of all worlds, one would predict in advance how large the difference to be compared was by conducting a pilot study or by reading other work. Once the likely variability of results is known, it is a simple statistical process to estimate how many samples are required. If the answer is hundreds more than could possibly be collected then it is pointless to continue. The study needs redesigning and a great deal of wasted effort can be saved. Similarly (though less commonly in practice) it would be possible to waste time in collecting more than enough data for the intended purpose. In this case the time could be more profitably used in developing other aspects of the study.

In truth, many bird census studies do not measure their precision at all and even a pragmatic aim of measuring it and trying to get within about 25% of the true value would be an improvement.

A still simpler point would be to urge the use of at least two plots before coming to any general conclusion. If their results are widely different then the conclusion cannot comfortably be generalised. Similar results from two plots could be due to chance, but even this low level of replication would well repay the extra effort in improving the confidence with which the results can be regarded. It is important that replication counts different birds in a different area. Apparently increasing the numbers of birds counted by making more visits is not the same thing.

In the example we are using, a critical feature is that we are trying to estimate the density of Skylarks in some fixed area which is larger than we could survey in full. If a full survey was possible because the fixed area was quite small, then questions of precision do not arise in the same way. On the other hand, the result could not be viewed as general and likely to apply elsewhere. It would merely say how many birds were on the area concerned, perhaps one farm.

We suggest that precision of the density estimate should be measured where possible by counting more than one area. How should the areas be selected? The answer is that they should be randomly located within the total area over which we want the answer to apply. The critical feature of randomness is that, as each plot is chosen, it should have equal probability of falling anywhere within the study area. Its location should not in any way be influenced by any prior knowledge which might bear on the expected number of birds. This requirement is essential to ensure that the resulting answers are representative of the study area without bias (see below).

Randomly located plots are picked from a numbered list of all possible plots that could be surveyed, using random numbers which are readily generated by computer, or looked up in a table. They should not be picked by eye in any way such as sticking pins into a map. The human eye is very bad at picking a set of random points. Still less should they be selected by looking around and trying to pick a range of places that seem to look pretty average for the area. An increasingly common practice is to use the national grid as a framework and to sample randomly selected 10 km or smaller squares. This makes picking random ones easy to do and is a useful practice in study design.

Although much of the variability between estimates from repeated samples is a property of the circumstances under study, there is another way of treating it. Variation between the sample plots might cause some of the spread of individual density estimates. It might be better to recognise the habitat variation and sample recognisable types separately. In the farmland Skylark example we perhaps ought to try to measure densities in different crops; taking some plots in sugar beet and others in wheat (Box 2.3). The same effort might then obtain more precise estimates for individual crops than for the wider more mixed area. Plots would be chosen with knowledge of the distribution and abundance of the various habitats, but the number to be chosen in each would depend on the objectives of the enquiry. In many cases it will be desirable to put more effort into the less frequent habitats than would be the case if the samples were taken at random. This sampling procedure is known as stratification. All possible plots or grid squares are attributed to a stratum (such as crop type) and the sample squares are picked randomly from within each stratum. The habitat variation may not be sharply split like crops but be continuous like the scrub cover on some grassland plots. In this case, plots could be spread along the range of variation, the scrub cover measured and the analysis performed by regression methods.

Accuracy

In seeking precision, we are dealing with random errors which on average tend to become zero with large samples. A different class of error is systematic in the sense that it goes in one direction and thus does not disappear with a large sample. This is known as bias. Methods free of bias are said to be accurate. In discussing the Skylark example with respect to precision we postulated a perfect census method with no bias. Imagine that in practice we wanted a good number of samples so chose a very quick method. This entailed standing in the middle of a plot and counting the maximum number of birds that were singing at any one time (Box 2.4). Obviously it is likely that this count will be larger in a place with more Skylarks. Equally obviously, it is most improbable that all the birds breeding in the plot will be singing simultaneously. Thus, unless birds from elsewhere or non-breeders are also coming in and being recorded, the count will not exceed the true number present but will inevitably tend to fall short of it. The results of this study would be said to be biased or inaccurate. They will on average fall below the true value but unfortunately we do not know to what extent this will be so.

Bias is inevitable in almost any realistic bird census method and in general we know neither how large it is nor in what direction it lies. It might arise from several sources, some of which exaggerate the result while others lower it. It is common, but not at all desirable, to list several possible sources of bias and suggest that they will cancel each other out and then ignore the problem. On the other hand, it is increasing awareness of countless sources of bias that has made much recent bird census literature so depressing. If, as is generally the case, accurate results are very difficult or perhaps nearly impossible to obtain, the perfectionist can argue that it is safest to give up.

It is possible to take a more pragmatic view. This entails being on the lookout for bias both in the field and in the literature with which one might be comparing results. Secondly, though much harder to deal with than random error, bias is not totally intractable. There are three kinds of action available. Firstly, if the likely sources are anticipated, steps can be taken to minimise bias for the particular project. Secondly, with careful design it is possible to sidestep the problem by confining comparisons to entities that have the same bias. Finally, it is possible, though often very difficult, to measure bias.

Bias is minimised by recognising its sources (see below) and not allowing them to creep into the study design. Counts are restricted to a fixed season and time of day. Observers are trained. Work is suspended if the wind is too strong. The counting methods are standardised so that point counts last for a fixed period or transects are walked at a fixed speed. Within a single study, the sources of bias that cannot be totally removed should be spread around fairly. If, for instance, there are ten plots each to be visited eight times, the order of visits should be designed so that the seasonal spread of visits to each is similar. This might be more expensive in travel but, if the first plot had most of its visits in April and the last was mainly counted in June, then bias would have been needlessly built into the results. On mapping plots, the direction of walking should be altered across visits. Otherwise one area will tend to be walked at a different time of day from another and spurious differences in bird distribution might be added. The Common Birds Census evades bias due to different observers by making year-to-year comparisons only where the same observers cover the same plots in a pair of years. Avoidance of bias can be thought about in the same way that experiments are designed. The purpose of an experiment is to isolate the factor(s) of interest from all other variables which could influence the outcome. The counting methods are just one class of such variables which should not be allowed to add unwanted variation likely to obscure the point of study.

Measuring bias is much more difficult. It can be done only if the true value can be ascertained. In practice this is somewhere between very difficult and completely unachievable. For a single species, it might be done by use of marked birds (Chapter 6), or by nest-finding and mapping on a small subset of study areas (Chapter 3). Different observers or more effort might be put into some study areas. If different observers come up with different answers then the possible scale of bias can be guessed, but not measured until an unbiased estimate has been obtained. For a community of birds, an unbiased estimate is probably not possible. If bias can be measured on a few plots, however, then it can be corrected on the other study areas counted by a quicker method.

In general, there is likely to be a trade-off between precision and accuracy in study design. The most accurate methods are the most expensive in time, so using them tends to preclude collecting many samples to maximise precision and increase the generality of results. It is our view that, in Britain, a tradition of using mapping methods (Chapter 3) for songbirds has represented an attempt to obtain better accuracy which may have been prejudicial to what might have been learnt by using quicker methods (Chapters 4 and 5) and collecting a greater spread of possibly less accurate data. The attempt to produce completely accurate results at the expense of leaving precision and generality unattended may often be futile. Future study designs will be aware of bias but will seek to optimise the trade-off with precision and generality (Box 2.5).

Sources of bias

In designing a study to handle bias it is necessary to understand its known sources. In many cases, a sensible response is then obvious.

1. The observer

Different people vary enormously in their birding skills and (in the case of some professionals, though probably not amateurs) their motivation. No-one would put much trust in comparing counts made across several years on a reserve where the work is the annual chore of a different, new and inexperienced assistant. At minimum, it is essential to be familiar with the study birds, including their calls and songs if appropriate. Deafness and failing eyesight may creep up with age. Beware. At a major bird census conference it was found that a large proportion of participants were partially deaf. It is also desirable to feel enthusiastic rather than run down by tiredness, cold or hunger because the schedule is too demanding.

In counts involving several people it is possible to spread the work in such a way that observers swap places, so results are less likely to be due to observer differences. This also serves as training by drawing attention to unusual results. More direct training may sometimes be desirable so that methods are used similarly between observers. This is especially the case if any difficult skills, such as counting large wader roosts or estimating distances, are involved. In general, we would always recommend that thought be given to training observers and ensuring that they meet a minimum standard before their results are used. If face-to-face training is not possible, then the use of full and clear written instructions can serve a similar function.

The mapping method (Chapter 3) has a special additional observer influence in interpreting the field results. The rules by which this is done are not sufficiently full for several people to be able to read and execute them and get the same results. BTO staff analysing the CBC are trained but this facility is obviously not available to anyone contemplating their own mapping study.

2. Census method

Different census methods undoubtedly vary in their susceptibility to bias, though this is rarely known. For this reason care should always be taken to state methods in written results, particularly if any special variations have been allowed on 'standard' methods. Putting effort into finding nests for instance can have a major effect on the results of a mapping census. It may make it more accurate, but it is different from the normal standard and thus leaves a bias in comparison with someone else's study. A more insidious way of adding bias could be the use of prior knowledge, for instance of the home patch. Imagine the effects this could have in atlas studies when comparing remote areas with those that are someone's well known local patch. Even if time spent in the field is similar, more of the different species actually present will be detected on the home patch because the observer knows what to expect and where.

Results obtained by different methods are very likely to have different biases. In general, if there are common rules for a method, they should be adhered to. Equipment is hard to standardise but could alter results. Someone with a good telescope is going to count more ducks than someone with poor binoculars.

There has been a tendency amongst songbird counters to believe that the mapping method is accurate and any other methods can have their accuracy compared with it. This view is now understood to be spurious. The fully accurate method is a very elusive thing. For the present, suffice it to say that comparison of any results derived by different methods will include bias which should be considered before biological conclusions are drawn.

3. Effort and speed

You generally see more the harder you try, either by walking slower or by putting in more overall time (Box 2.6). Within a study, effort should be standardised across years or plots or whatever. Size of plots might be considered as part of effort. If plots of differing size are involved, they should receive similar effort per unit area rather than similar total efforts. For common methods, there are normal standards which should be followed. The trade-offs with other considerations apply very obviously to effort. Consider the question of what would happen to the value of the results if the effort per plot was halved so that twice as many plots could be visited or vice versa. Unfortunately a magic formula for answering this question is not known. Intuition and expert judgement must play their part. One way of saving on effort so that it can be used for more plots would be to attempt to count fewer species. This is less heretical than it sounds once it is remembered that standard methods are so poor at counting some species (such as owls) that they are effectively not counted at all. So if some species are left out anyway, would it be possible to identify which are the critical species for a particular study and count only those? Would this make a valuable saving of time?

Effort is a particular problem in widespread surveys involving many people. If the ideal of making it constant is not achievable, then the next best thing is to measure it. In this way at least, it is possible to tell whether irregularities in distribution or numbers might be real or whether they were simply caused by variation in effort. Effort can also be recorded in any published results so that a reader comparing different studies can tell the extent to which their differences might simply have been due to differences in effort. In some analyses, the effects of variation in effort might be allowed for by statistical means. Related to recording effort is the recording of zero counts so that these can be treated as different from no records due to no effort. Birds of prey, for example, are often counted by checking previously known nest areas. In recording effort, both the reporting of zero counts and the effort put into intervening areas need to be considered.

4. Habitat

Birds are easier to find in some habitats than others. Some methods deal better with this potential problem than others. If it is not dealt with, it should be remembered as a possible source of bias. Bibby and Buckland (1987) show how birds are more detectable in open habitats than scrubby ones and how this can be allowed for in estimating densities from point counts (Box 2.7). Barn Owls are quite readily found if they nest in barns which can all be located and checked in a study area. If they nest in trees, they will be much harder to find. Since the use of these two kinds of sites varies regionally (Shawyer 1987), there is considerable risk of bias in comparing results from different parts of Britain. A well designed study would need to recognise this bias arising from habitat differences and deal with it by using sufficient effort to find tree nesters.

Some habitats such as dense scrub or marshlands are difficult to count in because of sheer inaccessibility. Densities of birds in suburbia are poorly known because of a different kind of inaccessibility. Near rivers, roads or industrial sites, there may be so much noise that quiet bird calls or songs are very difficult to detect. Even outside the breeding season, sound is more important than sight for detecting small birds in thick vegetation. It may simply not be possible to obtain accurate counts in some circumstances.

5. Bird species

Different bird species also vary in their susceptibility to being counted. Some are noisier than others (Box 2.8). Some breed late, some early. Some are readily mist-netted, others not. It is most unlikely that a generalised method will count all species in similar units. It might therefore be valid to compare results within a species but not necessarily between species. No general methods work on all species, and some require special methods (Chapter 7). In some species, the counting methods might count only parts of the population. Counts of breeding seabirds or birds of prey can include only the breeding part of the population. In long-lived species with delayed maturity, the pre-breeders may be a sizeable part of the total population and may escape counting. This would be a serious bias if, as is plausible, a population was declining but this allowed immatures space to breed earlier. In such circumstances, a decline might not be noticed because it was primarily the pool of pre-breeders that was declining while the counts were just of the breeders. The extent of this possible scenario is poorly known and may be larger than generally appreciated even in short-lived passerines.

6. Bird density

At high densities, the observer may be swamped by the problem of recognising different species or individuals or territory boundaries (Box 2.9). At low densities, boredom may tempt lack of thoroughness in searching. Counting very abundant birds poses special problems considered later (Chapter 8). Counting very scarce or dispersed species returns to the need to ensure that effort is properly distributed and recorded.

7. Bird activity

Individuals may vary so much in their detectability according to their activity as to make comparison difficult or impossible. Counting Puffins sitting at a colony would, for instance, say little about the numbers on nests underground or out of sight at sea. The occurrence of the activity that makes birds countable may itself be related to weather, time of day or time of year. Feeding waders may be dispersed over a wide area and conspicuous. When roosting, the same birds may be in a tight flock which could be completely overlooked. It is in general very difficult to count breeding birds in a comparable way to counts at other seasons. In some cases, counting methods may be deliberately aimed at part of the population involved in a particular activity (Chapter 7). Breeding duck numbers, for instance, are often assessed by counting the gatherings of males in the early part of the breeding season.

Colonial birds are often counted in a way that is related to breeding success. The number of birds in attendance at a colony could well be related to whether breeding success has been good or poor that year.

Numbers of Arctic Terns breeding in Shetland appear to have declined, as well they might after several years of very poor breeding success. How confident can we be that the decline is properly estimated? It is quite possible that food supplies are so poor that many individuals do not attempt to breed and, together with those that have lost eggs or chicks, they might stay at sea and avoid being counted in colonies. Green and Hirons (1988) provide a model which shows the effect that such a phenomenon could have on such counts. In general, counters should be aware of this possible bias and consider measuring breeding success so that possible effects of its variation are not overlooked.

8. Season

Breeding birds, especially, vary in detectability by season (Box 2.10). The best period for repeatable counts may be very brief. Many warblers, for instance, sing for rather few days and then become much quieter once mated. At the same time, growth of vegetation can rapidly make counting harder in the early summer. Count periods therefore need to be carefully standardised for comparison. Across years, the standardisation should ideally be by the birds' season rather than the calendar. Unfortunately, there is a wide difference between species in the best period for counting so a general census method may involve compromise.

9. Time of day

Time of day should similarly be standardised because of variation of activity (Box 2.11). The greatest output of song for some species is close to dawn and is sometimes so vigorous as to be overwhelming. Moreover, the rate of change of singing intensity near dawn is often high. Common advice is therefore to count songbirds starting shortly after dawn and to stop by mid-morning. Again, standardisation is important.

10. Weather

Extremes of weather affect bird activity and the comfort and acuity of observers (Box 2.12). High wind-speeds pose the greatest difficulties for songbirds which are harder to see or hear if trees are moving and noisy. Light rain, in contrast, is rarely a problem. Morning activity is often terminated earlier in hot weather. For long-range counting of wildfowl, waders or seabirds, light intensity and visibility are important. The best advice is to avoid counting in poor weather. It is difficult to specify just what this means, especially as weather factors are often related to each other and to time of day or year. In some circumstances, it might be possible to measure the weather and allow for it but this is rarely fruitful.

Summary and points to consider

1. Precision

What is the total area under study?
What kind of study method is required?
Would a more accurate/less accurate approach with smaller/bigger sample sizes be better?
Is it necessary to count many species or would fewer do?
How are plots/points/routes to be distributed?
Are they representative of the area being studied?
Would a stratified design be better?
Are there enough plots/points/routes to get a sufficiently precise answer?

2. Accuracy

What steps are to be taken to deal with bias from the following?
    Observers
    Methods
    Effort and speed

Habitat
Bird species
Bird density
Bird activity
Season
Time of day
Weather

Can some bias be eliminated?
Can remaining bias be spread similarly across all plots?
Can bias be measured?
Should any other factors, which might cause bias, be measured?

3

Territory Mapping Methods

Introduction

During the breeding season, many species are territorial. Especially among passerines, territories are often marked by conspicuous song, display and periodic disputes with neighbours. Often, the area is not completely filled with territories because of low densities or gaps in suitable habitat. In such cases, mapped registrations of birds should fall into clusters approximately coinciding with territories. Where a species has closely packed territories, the mapping of simultaneously singing birds presumed to be in their respective ranges becomes important. Territory boundaries are taken to be between such birds. The mapping approach relies on locating all these signs on a series of visits and using them to estimate locations and numbers of clusters or territories.

This method has formed the basis of the BTO's Common Birds Census since 1962 (e.g. Williamson 1964) and has been widely used elsewhere (William 1936; Kendeigh 1944; Enemar 1959). To some people, it has come to be seen as the standard against which other methods can be compared. This view is not justified. Mapping does not work well on birds that do not show much territorial behaviour, especially semi-colonial species and those that do not sing. Even among passerines there are troublesome variations. Many migratory warblers sing for a brief time before finding a mate and becoming inconspicuous. Some species, such as Linnets, nest in loose aggregations with little territoriality. Other species such as Pied Flycatchers or Wood Warblers may sing in more than one territory and keep quiet while moving between the two. Some species, such as Reed Warblers, occur at high densities but can move within a season if successive nests fail.

Most of these problems do not matter if the methods are standardised and the results are used for population-indexing as in the CBC. In such an instance, standardisation of method is critical. There are rules for mapping census work set out by the International Bird Census Committee (International Bird Census Committee 1969). As discussed later, the rules are a matter of some controversy, and the methods needed to make mapped results more accurate go beyond the rules. In attempting to obtain good density figures or meet a purpose not reliant on comparison with other studies, it might be better to ignore the constraint of these rules.

The mapping method is the most time consuming of the general bird count methods for a fixed number of birds finally counted. In this sense, it is inefficient. For this reason, it is often difficult to design studies requiring the representation of a range of habitats or with experimental treatments, controls or replication. Point counts or transects might be considered instead in such cases. The mapping method really gains if use is made of the fact that the data are mapped, i.e. if inferences are drawn about the relationships between the distribution of birds and of habitats.

Field methods

1. The study plot

The location of plots needs careful consideration depending on the objective of the study. Are the results going to be claimed to have any generality over and above describing what is actually on particular plots? If they are to describe the farmland birds in a particular area, then they need to be selected in some systematic way (see Chapter 1). Are the plots big enough to encompass any scarcer species of particular interest?

A study plot needs to be adequately mapped at a scale of about 1:2500 (Box 3.1). It is important that birds are mapped accurately and, at this scale, symbols can be positioned to within about 10–20 m which is probably comparable to the error to be expected in map reading. Other scales between 1:5000 and 1:1250 may be preferred depending on the density of birds on the plot. It may be useful to reproduce copies from a traced and simplified outline if the map is too cluttered. In open or uniform areas, it is essential to mark selected stones or trees, or other features on the ground and on the map so that any bird can be located accurately. In many woods, it is necessary to survey the plot extensively in advance to make sure that it is possible to map the birds accurately. In woods, it may become very hard to see far at the height of summer.

In woodlands with high bird densities, a plot of about 10–20 ha is suitable for coverage in a single visit of 3–4 hours. On farmland, about 50–100 ha can be covered depending on the number of hedges and woody areas. It is necessary to walk the boundaries of the plot, so on arable farmland it is recommended that these should be field boundaries. This approach will exaggerate the average density of birds on farmland, since the bulk of the birds are in the hedges. It is particularly undesirable to choose boundaries that contain a lot of birds such as those against a woodland. Because of the problem of edge effects, it is better to have plots that are roughly square or round and to avoid those with long and complicated edges. The edge effects become relatively less important the larger a plot becomes. An upper limit will be set by the time taken to cover the area properly.

To make the most of a mapped plot, it is necessary to describe its vegetation in some way (Chapter 10). Since the bird data are mapped, it is ideal to map the vegetation as well. This allows a variety of further analyses on habitat utilisation. If the plots are each relatively uniform and there are several of them that differ from one another, it might be sufficient to describe or measure the vegetation without mapping it.

2. Time and route of visits

The results of a mapping census can be influenced by the number of visits. It might be thought that more visits would be better, but in fact they can add confusion rather than clarity. The CBC has adopted ten visits as a standard. In southern English woodland, these would take place between about mid March and mid June. Ideally they would be spread fairly uniformly at about weekly intervals. For any one species, all ten visits are rarely needed. The total number of visits and the length of the season should give all species enough registrations to clarify clusters. The important visits for most resident species will fall in the first half of the survey. Later arriving migrants will not be recorded until the later half of the season. By then, many of the residents have young, and territories are neither so important nor so clearly advertised and defended.

Early morning is the best time for visiting, but some (in the CBC up to two) evening visits might be helpful. It is best to avoid the first hour of activity before dawn. At this time, bird activity peaks very markedly, so there is a risk that the part of a plot covered first will produce more records. A period of more uniform activity lasts from about sunrise to about midday. On hot days, these periods may be briefer. Since time of day and differences in effort can cause bias, it is important to record data and start and finish times as part of the documentation of a visit and to standardise these variables as far as possible.

The plot should be walked at a slow pace so that all birds detected can be identified and located. The route should approach to within 50 m of every point on the plot (Box 3.1). In thicker vegetation, a closer approach would be better. On farmland, all hedgerows usually need to be walked. Routes and directions should vary between visits so that there is no systematic tendency for any particular part of the plot to be visited later or earlier in the day, on average. Single visits should be completed in a single period of fieldwork. Splitting visits across several days may cause problems with double recording of the same birds. If visits are split, it is useful to record them as such. Woodland can be surveyed at the rate of about 5 ha per hour, while farmland might be covered at 20 ha per hour. The duration of a visit depends on bird activity, which may be good for up to 6 hours, and on the stamina of the observer.

3. Bird recording

The identity and activity of all birds are mapped with small and tidy writing in pencil or ball-point pen. If the map gets wet, some inks run; ball-points do not work on a wet map but pencils still do. It is helpful to use a standard list of codes for bird species. Codes for some common British birds are shown in

Base maps for territory mapping.   **Box 3.1**

(a)
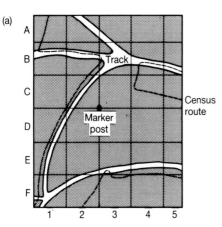

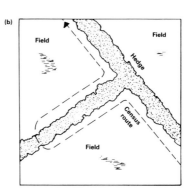

(a) Part of an outline map for a woodland plot with a 20 m grid. Grid-points are marked A1, B2, C3, C4, etc. The map also shows tracks, habitats, grids, marker posts or other features. Observers can follow the grid-lines, using a compass if necessary, so that they always know their position on the map.

(b) For studies on more open farmland, the use of the grid is unnecessary as topographical features such as telegraph poles, trees, stakes hammered into the ground at known positions, houses, etc. can be used to fix the position of the observer, and hence ensure accuracy in mapping of the registrations. In this example more birds are expected in the hedge than in the crops, and it is often difficult to walk through crops, hence the census route follows a line closer to the hedge than the centre of the field.

Box 3.2. Activities are also given standard codes (Box 3.3). Care should be taken to record as much detail as possible such as the sex and age of the bird. It is often critical in analysis to know whether a multiple observation was a party of juveniles. Were two different birds nearby the male and female of a pair, or were they of the same sex and thus presumably near a territory boundary?

The most useful point to concentrate on is the location of individuals of the same species that can be seen or heard simultaneously. A key feature of analysis is the assumption that territory boundaries fall between such records. For uniformly distributed species, it is difficult to analyse results without such simultaneous registrations. Is the bird now singing different from the one seen 50 m away and 2 hours previously? The ambiguous registrations must be linked with a question mark to indicate uncertainty. They are quite likely to be the same bird. If they really are different, the chances are that both will appear simultaneously on a different visit. Searching for nests is not a good use of time on mapping surveys, though the information is used in analysis if available.

Box 3.2

Standard codes for mapping birds.

A selection of standard species codes developed by the British Trust for Ornithology for British passerine species commonly recorded on Common Birds Census plots. The full list (British Birds 1984) can be used for any kind of bird recording in all habitats and conditions throughout the Western Palaearctic.

SD	Stock Dove	MT	Marsh Tit
GS	Great Spotted Woodpecker	WT	Willow Tit
S	Skylark	CT	Coal Tit
TP	Tree Pipit	BT	Blue Tit
WR	Wren	GT	Great Tit
R	Robin	NH	Nuthatch
N	Nightingale	TC	Treecreeper
SC	Stonechat	MG	Magpie
B	Blackbird	RO	Rook
ST	Song Thrush	C	Carrion Crow
SW	Sedge Warbler	RN	Raven
RW	Reed Warbler	SG	Starling
LW	Lesser Whitethroat	CH	Chaffinch
WH	Whitethroat	GR	Greenfinch
GW	Garden Warbler	GO	Goldfinch
BC	Blackcap	LI	Linnet
WO	Wood Warbler	LR	Redpoll
CC	Chiffchaff	BF	Bullfinch
GC	Goldcrest	Y	Yellowhammer
PF	Pied Flycatcher	RB	Reed Bunting
LT	Long-tailed Tit	CB	Corn Bunting

Standard symbols for bird activities.   **Box 3.3**

The standard BTO list of conventions is shown. These are designed for clear and unambiguous recording. Symbols can be combined where necessary. Additional activities of territorial significance, such as displaying or mating, should be noted using an appropriate clear abbreviation.

CH, CH♂, CH♀ 3CH juvs, CH♂l♀	Chaffinch sight records, with age, sex or number of birds if appropriate. CH⚥ indicates one pair; 2CH⚥ means two pairs together.
R fam	Juvenile Robins with parents(s) in attendance.
R	A calling Robin
R	A Robin repeatedly giving alarm calls or other vocalisations (not song) thought to have strong territorial significance.
Ⓡ	A Robin in song
RR	An aggressive encounter between two Robins.
* R	An occupied nest of Robins; do not mark unoccupied nests, which are of no territorial significance by themselves.
⊠ BT	Blue Tits nesting in a specially provided site (e.g. nestbox)
* PW on	Pied Wagtail nest with an adult sitting.
PW mat	Pied Wagtail carrying nest material
PW food	Pied Wagtail carrying food.

Movements of birds can be indicated using the following conventions:

— GR →	A calling Greenfinch flying over (seen only in flight)
Ⓓ →	A singing Dunnock perched then flying away (not seen to land)
→ B♂	A male Blackbird flying in and landing (first seen in flight)

(Continued)

**Box
3.3
*cont.***

The following conventions indicate when registrations relate to different birds, and when to the same bird. Their proper use is essential for the accurate assessment of clusters.

WR ⟶ WR

A Wren moving between two perches. The solid line indicates it was definitely the same bird.

Two Wrens in song at the same time, i.e. definitely different birds. The dotted line indicates a simultaneous registration and is of very great value in separating territories.

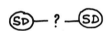

Two Linnet nests occupied simultaneously and thus belonging to different pairs. This is another example of the value of dotted lines. Only adjacent nests need be marked in this way.

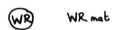

The solid line indicates that the registrations definitely refer to the same bird.

(SD)— ? —(SD)

A question-marked solid line indicates that the registrations probably relate to the same bird. This convention is of particular use when the census route returns to an area already covered—it is possible to mark new positions of (probably the same) birds recorded before, without the risk of double recording. If birds are recorded without using the question-marked solid line, overestimation of territories will result.

(WR)   WR mat

No line joining the registrations indicates that the birds are probably different, but depending on the pattern of other registrations they may be treated as if only one bird was involved. (It is possible to use a question-marked dotted line, indicating that the registrations were almost certainly of different birds.)

C*   C*

Where adjacent nests are marked without a line, it will often be assumed that they were first and second broods, or a replacement nest following an earlier failure.

In all cases the standard BTO codes for British birds should be used.

A field map.

Box
3.4

Part of a completed visit map for a woodland census, as used in the field (from Marchant 1983). This is one of the ten visits recommended for such work, spread over the spring breeding season and conducted in the early to mid morning. Such visits are labelled as A to J, and all the registrations of birds are presented in the standard way. Other important factors such as the weather, wind force (W3), date of the survey and observer name should all be appended to the survey map so that detailed comparisons between sites in the same year or the same site in different years can be made with the knowledge that other variables are not influencing the data.

In this instance it was a productive visit and all parts of the map are crowded with registrations from different species. The dotted lines will be particularly helpful in the later analysis of territories. Blackbird registrations have already been copied to the species map and cancelled with a light stroke of the pen.

A species map.

Box
3.5

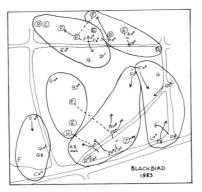

This is the Blackbird species map from the same census as in Box 3.4. On transfer to the species map the B for Blackbird has been replaced by the visit letters A–J which represent the ten visits. However, the symbols indicating sex, song and movements have not been changed. The map has already been analysed, and six territories were found in this portion of the plot.

Box 3.6

Minimum requirements of a cluster (from Marchant 1983).

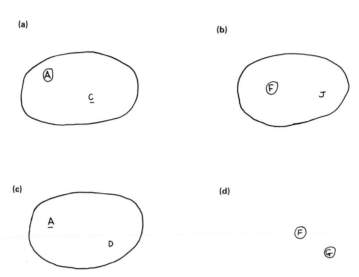

(a) There are two records of a Robin on visits A and C. If there had been only eight census visits to this plot then this amount of data would be sufficient to define a territory; however, if there were nine or the recommended ten visits to the plot then the two records would be insufficient to define a territory.

(b) There are two records of Whitethroat on visits F and J (early May and late May) out of a sequence of ten visits. Because Whitethroat is a migrant which does not arrive in Britain until relatively late in the breeding season these two registrations are sufficient to mark this as a breeding territory.

(c) There are two records of Tawny Owl on visits A and D. These are sufficient to define this as a breeding territory because Tawny Owls are difficult to count and hence the minimum requirements for a territory are lowered.

(d) There are two records of Willow Warbler on visits F and G. However, these are separated by only 2 days, not the required 10, hence the territory is not valid as the records may well refer to a bird that sang briefly whilst on passage to another site and did not breed.

Dotted, solid and question-marked solid lines.

Box 3.7

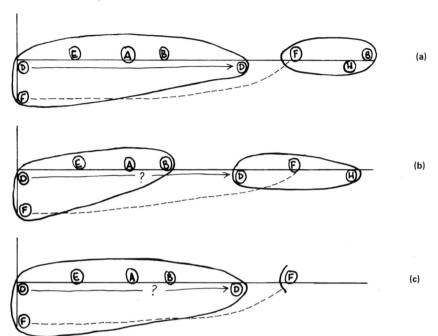

This figure shows examples of the correct treatment of lines between registrations of Willow Warblers (from Marchant 1983).

(a) The dotted line between the two warblers recorded on visit F means that the two F registrations cannot be placed in the same cluster. The solid line between the two birds recorded on visit D means that both records were of the same bird and should be placed in the same cluster.

(b) and (c) The question-marked solid line between the two birds recorded on visit D can be treated in either of these two ways, depending on the pattern of other registrations. In (b), there are sufficient registrations to support a second cluster *DFH* and the D records are treated as being of separate birds. In (c), there is no support for a second cluster because there are fewer than two birds recorded in the second possible cluster, and hence the D records are treated as if one bird was involved. In this case the second F registration is treated as a superfluous registration.

These examples are correct as they stand, but on a real map they might be further influenced by the pattern of adjoining registrations.

Box 3.8

Interpretation of dotted lines.

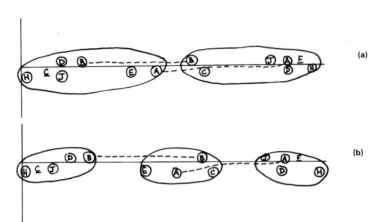

(a)

(b)

In (a) and (b) there are two different analyses of the same set of registrations of Willow Warbler (from Marchant 1983). Standard BTO instructions to counters state that example (a) is unsatisfactory because the apparent nucleus of registrations on visits *ABC* is split between two clusters. Example (b), giving three smaller clusters, is said to be a better analysis because it uses *ABC* as the basis of a separate cluster. The treatment of dotted lines is correct in both examples.

Box 3.9

Large diffuse clusters.

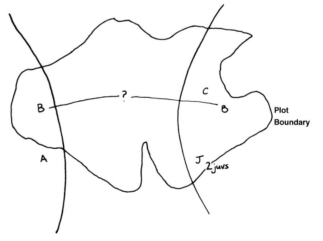

When mapping large and mobile species such as Kestrel, it may be difficult to gain sufficient records to be sure of territory boundaries. In this example either one or two birds were seen on visit B. It was not known whether they were the same bird or not. Other records give the impression of two clusters and the interpretation of the map is the best fit of the available data. Only one territory is confirmed.

Territory shifts and multiple registrations.

Box 3.10

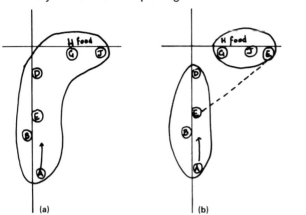

Three correct examples of analysis are shown (from Marchant 1983).
(a) The two groupings *ABDE* and *GHJ* are merged into a single cluster on the assumption that there has been a shift of territory. It would be wrong to draw clusters where such groupings are so close together.
(b) The addition of a second E and a dotted line makes it clear that there are two clusters.
(c) The figure is extended to show a correct treatment of multiple registrations. Neither cluster has any double registrations.

Semi-colonial species.

Box 3.11

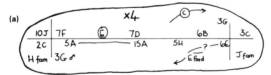

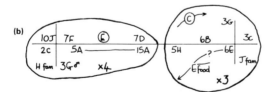

For a semi-colonial species such as Linnet it is often necessary to draw clusters representing groups of territories. Examples (a) and (b) show correct and incorrect treatments of the same set of registrations. Example (a) is correct, based on totals of seven birds on visits D, E and F. The high count on visit A is discarded as probably a remnant of a winter flock, while that on visit J probably includes juveniles. Example (b) is incorrect, since the peak counts in these two adjacent putative clusters occurred on different visits, and combining them as in example (a) considerably reduced the assessment (from Marchant 1983).

Box 3.12

Edge clusters.

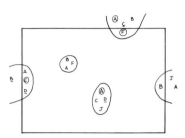

In this example of a territory map there are three possible interpretations of the number of territories depending on the system being followed. In the CBC (Marchant 1983) all the edge territories would be counted as well as those wholly within the plot; this method would therefore give five territories for the plot. In the International Bird Census Committee methodology (International Bird Census Committee 1969), edge clusters are treated differently and only those clusters in which most of the registrations occur within the plot are counted as bird-territories for the plot. In this case there would be three territories within the plot. In an alternative system only the territories wholly within the plot are counted as full territories, others are accorded a proportion. In this case the number of territories would be interpreted as c. 2.5 territories.

That the CBC regards all edge territories as a part of the total number of territories for the plot can lead to an overestimation of the density of birds on the plot unless allowance is made. Errors are worse if the margins are rich in birds. It is recommended that plots are selected so that their margins are typical of the plot in terms of their bird abundance in order to minimise this effect.

Box 3.13

Population indices for Song Thrush and Blackcap derived from CBC data in British woodland and farmland habitats.

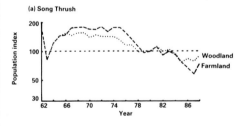
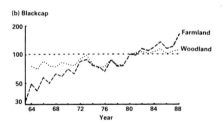

(a) Song Thrush. This graph indicates population trends caused by both weather and other factors. The harsh winters of 1961/62 and 1962/63 caused populations of this resident British species to decline markedly. However, since the mid 1970s the population has been in longer-term decline, due in part to cold winters in 1978/79 and 1981/82 but also to unexplained factors (from Marchant *et al.* 1990).

(b) Blackcap. The CBC index for this migratory species shows a long-term increase throughout the period 1963–1988. There are various theories for this increase which are summarised in Marchant *et al.* (1990) but, as those authors conclude, no reasons can definitely be ascribed for this increase.

Some territories will overlap the plot boundary. Several ways of dealing with these are considered below. Generally it is necessary to map records outside the plot. In the field, this is usually accomplished by recording everything detected from the boundary without walking outside the plot. The field maps therefore need to extend by about 100 m in all directions beyond the plot.

Variations of the method

Mapping is the method used in the BTO's Common Birds Census. In this case, it is vital that the method does not vary. For other uses, however, there are ways of making mapping more accurate and these might be considered.

1. Restricting the list of species covered

The CBC and bird community studies include all species, or at least as many as can possibly be recorded. If the purpose of the study is more restricted, the first variation is to map fewer species. This helps in concentrating time and effort in the required direction. The purpose of making ten visits is to gather enough data to register territories of species differing in their breeding and singing seasons. If only one species is being studied, five visits at the most appropriate season and time of day for the species would suffice.

Recognising adjoining birds as different is the single most important part of the field data for the support of the analysis. In a study aimed at a particular species, more effort can be put into finding neighbours singing at each other and marking their positions while they do so. The species map could be updated after each field visit. In this way, it would be obvious if there were any areas where it was going to be difficult to resolve the number and pattern of clusters. Extra field effort could be made in these places on subsequent visits.

2. Eliciting responses

The observer can increase the chance of finding birds and obtaining simultaneous registrations by using a tape to play snatches of song of the target species and recording any responses. If a tape is played in places that may be territory boundaries, it might help to see whether or not there is a response from both birds. Owners of isolated territories often sing less than those where the neighbours are close. A tape recorder could increase the chance of getting sufficient records for a cluster. Many migrants only sing early in the breeding season, but a tape recorder can get a response while birds are nesting and have largely stopped singing. Thought should be given, however, to the risk of disturbance caused by this method.

For some elusive birds, dogs can be used to increase the chance of obtaining registrations. Dogs are regularly used for counting gamebirds and for finding nests of ducks or waders. It goes without saying that a suitable and well trained dog is needed if it is to do more good for the census than harm to the birds.

3. Consecutive flush

Another way to make birds respond is to try to chase them to their territory boundaries. They will often fly readily within the territory but be reluctant to go beyond it. In this way, the extent of an individual's territory might be sketched with 10–20 flushes trying to move it round its boundary. With any luck, the neighbours might be seen or heard responding at a territory boundary. Critical areas for interpretation can be tested again if the species map is updated after each field session. Fieldwork need continue only until a coherent result has been reached with all clusters meeting the minimum requirements and no spurious registrations unexplained.

4. Nest-finding

The disadvantage of looking for nests in a study involving many species is that it is unlikely that they can all be found with equal ease. In a study of a single species, the location of nests may be needed for other purposes. Since they must belong to a pair, they provide the strongest possible evidence of the presence and location of that pair. Often the evidence from the distribution and timing of nests will be consistent with that from the more conventional mapping.

For some species with weak signs of territorial behaviour, nest-finding is the only really good way of counting them. Some such as corvids or pigeons have fairly conspicuous nests that are quite easy to find. Other nests such as those of ducks or waders are much harder to find. The disadvantage of using nest-finding alone is that it is hard to know when all the nests have been found. There is a risk that the more easily found ones might be a biased sample. If mapping is used as well, there is the possibility of continuing field-work until the results of the two approaches are consistent.

5. Marked birds

Much of the difficulty in interpreting maps comes from the problem of knowing which records refer to the same individual and which to different ones. This problem becomes much simpler if birds are uniquely colour-marked or radio-tagged. Territories can be drawn from records of known individuals. A drawback of this approach is that it is very time-consuming. The birds have to be caught and marked, and more effort is needed in their field recording. Colour-rings can often be diffcult to see and read, and radio-tracking is time-consuming. Marking is discussed in Chapter 6.

Marking is the only way of dealing with interpretation of those individuals in the population that are not breeding and are therefore behaving in a way incompatible with the assumptions of the territory mapping method. As more studies are published, it is becoming increasingly clear that many populations, even of short-lived songbirds, contain a sizeable floating population of non-breeders.

6. The full study

If marking and nest-finding are used, and all efforts made to maximise the critical records for a mapping study, there is a chance that the results will give a good picture of the absolute number of birds using an area in a season. The output will of course include much more detail than a mere count. Such studies represent the only known way to obtain absolute counts for many birds. Needless to say, they are rarely attempted because of the effort required.

Interpretation of results

1. Transposing the field data

Field maps are generated for each visit and contain all records of all species (Box 3.4). These have to be converted into species maps. The visits are identified by letter (A, B, C, D, etc). For any one species, registrations are copied from all the field maps to a single species map. The species code on the field map is replaced by a visit code on the species map (Box 3.5). It is difficult to extract all the records unless those transcribed are crossed through on the visit maps. Even then, it is worth a thorough check to see that all have been extracted. In practice, it is possible to plot several different species on a single sheet by the use of colours and selection of common and rare species or a pair that occupy different habitats and thus have little overlap. Ball-point pens are best: some coloured felt-tips fade rather rapidly.

2. Interpreting the species maps

Interpretations of maps is not such a simple process that unambiguous rules can be described. The BTO has published some general guidelines, which form the basis of this section (Marchant 1983). There may be more than one 'correct' way to interpret a particular map. For the CBC, the BTO analyses all species maps sent in by the field workers. Analysts can confer on difficult specifics and are trained to a consistent standard. For the purpose of the CBC, consistency is more important than any notion of absolute correctness. The main problem with the guidelines is that they include some circularity in calling for avoiding clusters that are too large or too small for the species. Unfortunately, the normal range of territory sizes assessed from an independent method is not known for most birds.

The general aim is to draw non-overlapping rings around clusters of registrations that refer to one pair of breeding birds. These clusters are often called territories and results are presented in terms of number of territories. It should not be thought that the rings drawn are necessarily territory boundaries or indeed say very much about where the birds might range in the course of the summer. They are essentially products of the analytical method.

In perfect circumstances, a species map would show a group of distinct clusters each of which would contain records of one or two birds on most or at least several visits. There would be several records of the male singing. Dotted lines would radiate to adjacent clusters to show that these had been seen in the field as belonging to two separate pairs. Some species cluster well and others do not (Fuller and Marchant 1985). In practice there will be some occasions when more than one registration in an apparent cluster came from a single visit. Was this a double recording of one bird or a temporary intrusion by another, or does it indicate that the apparent cluster in fact refers to two pairs? There will be other areas of the map with rather sparse registrations. Were these casual records of non-breeding birds or those briefly out of their normal ranges or are they the signs of a territory where the birds were normally missed for some reason? All clusters that meet a minimum standard for acceptance can be sketched in. Use a soft pencil so that changes can be made if necessary.

3. Minimum requirements for a cluster (Box 3.6)

There must be at least two registrations if there were eight or fewer effective visits for the species, and at least three registrations for nine or more effective visits. For migrants the number of effective visits is the number from the first visit on which the species was detected. If ten visits at weekly intervals started in late March, many migrants will not be seen at all in the first two visits. In this case they will have eight or fewer effective visits. They need only two registrations per cluster. Difficult species, especially nocturnal ones such as owls or Woodcock, are considered to have rather few effective visits and clusters can be counted from two registrations.

The records in a cluster must additonally be at least 10 days apart. This is to avoid including temporary migrants present for a few days and registered twice. This rule has to be dropped, however, if, as is sometimes done, mapped plots in remote areas are visited at a higher than normal rate over a shorter than usual season.

A single record of a nest with eggs or young can be counted as a cluster even in the event of the adults not having been seen at a level to qualify. Broods of flying juveniles or of nidifugous species such as gamebirds or waders should not be counted in the same way as a nest. They might have moved from a territory already recorded or from one outside the plot.

4. Dotted and solid lines (Boxes 3.7 and 3.8)

A dotted line indicating two different birds seen simultaneously should not normally be included within one cluster. The only exception can be where there is reason to believe that these birds are the two adults of a pair or an adult and juveniles. Two records joined by a solid line should not be split into two clusters, as they belong to the same bird. Uncertain records joined with a question mark can be treated either way in accordance with the sense of the other information.

It is the dotted lines that make it possible to analyse complicated maps. This is why their detection in the field is so important. Analysis can begin by sketching in the places that have to be boundaries between clusters because they are crossed by dotted lines. It is then possible to start completing clusters that contain the minimum numbers of records to be acceptable.

5. Multiple sightings

Records of more than two birds, other than mates or juveniles, together should generally be treated as belonging to more than one cluster. There is a legitimate cause of multiple observations if a single bird is unwittingly recorded twice. This is where careful field observation helps. If a group of birds showed any sign of aggression to each other, then it would be reasonable to put them on a boundary between two clusters. If during field-work, two records fall very close together on a single visit, it is worth another few minutes of waiting to see whether or not they are two different birds. This might be done by seeing whether the one still visible will fly past the place where the first was recorded, and if so whether there is a reaction by another bird. In some cases, multiple records will be due to migrants such as Blackbirds, which are still present in Britain during April while the local birds are breeding.

If a cluster contains multiple registrations on more than two dates or if they have any suggestive spatial division, it might be best to divide the cluster into two. Consider whether the division would yield two clusters which both meet the minimum standards for acceptance. Would they be of realistic size for territories of the species in question based on knowledge from nearby on the map? Some species are particularly good at moving rapidly and undetected across their territories. Examples are Chiffchaff, Wren and *Sylvia* warblers. Some migrants pause briefly and sing before acquiring a territory in the spring. Extraneous records of species such as Sedge or Willow Warbler might be due to this and do not merit a split of a cluster. Some species, such as Blackbird and Yellowhammer, move widely outside their territories or do not have very marked territories at all.

6. Superfluous registrations

Some records will neither be enough for a cluster on their own nor obviously be part of a nearby one. They should be included with the nearest cluster if this will not make that cluster too large for the species and will not cause it to have too many multiple registrations. Otherwise they should be left out. Some such records might belong to a territory largely located off the plot. In other cases, they will belong to floating birds, especially early in the season or later on if they are juveniles. The analysed map should include some notation of how such records have been treated to show that they have been considered rather than overlooked.

7. Large or diffuse clusters (Box 3.9)

In uniform areas and for some difficult species, records may be widely spread rather than grouped. It is generally best to start by drawing in the obvious possible clusters with guidance from groups of registrations or dotted lines. Then work from these centres to see how the rest of the data can be made to meet the rules. It is not usually helpful to start at one edge of a map and work systematically across it.

Large species may show diffuse clusters and are more likely to have territories that reach well beyond the boundaries of the plot. The best that can be done for these is to see whether there are any signs that parts of two territories meet somewhere on the plot. In cases of large or diffuse records, the clusters drawn should be appropriate for the size of territory known to be occupied by the species in question. The difficulty here is that this may not be known and may vary with population density or habitat.

8. Spurious clusters (Box 3.10)

Adjacent clusters may meet the minimum rules for qualifying but should be examined to see whether they could better be classed as one larger cluster. Joining the two could produce just one cluster that did not exceed the allowed number of double registrations. This should be preferred if it does not produce an unrealistically large cluster. It is likely that the pattern of field records arose from the birds having two preferred areas of activity with something of a gap in between. They might, for instance, sing in two separate patches on scrub and not in the intervening grassland. Another possible explanation might be that a territory has shifted during the course of the season. In this case, the early registrations will tend to be in one part and the later ones in another. Such cases should always be assumed to belong to a single pair if they still obey the rules when joined up.

Attention should be given to the possibility that very small clusters in fact better belong with a nearby group. Other spurious clusters may occur in preferred feeding areas. Be especially wary of those that do not have records of singing birds.

9. Colonial or non-territorial species (Box 3.11)

The mapping method really only works well for species that show clear clusters but is used on other species as well. For species that defend small areas near a nest-site or a female, but range widely, group clusters are drawn. Such species include hirundines, pigeons, ducks and some finches. The CBC uses the following guidance. The clusters must include a potential nest-site, so swallows feeding over a field would not be included unless there was also a building. If adjacent group clusters contain similar maximum numbers of birds on different visits they should be amalgamated. Group clusters should be large enough for the numbers of pairs attributed to them.

Each group cluster is given a number of pairs. This should be the highest or second highest count of males obtained on any one visit. The second highest is arbitrarily used to reduce the possibility of a spuriously high temporary figure for one visit. If the sex of the birds is not or cannot be recorded, they are assumed to be equally divided by sex. Excessively high counts, possibly due to temporary feeding flocks, are not included, so if one count is very high, then it is the third highest count that is actually used. Care has to be taken to avoid high counts in the spring before winter visitors or passage migrants have left. Similar care has to be taken to avoid young or post-breeding concentrations of species such as Lapwing.

If the number of nests or broods simultaneously recorded in a cluster is larger than the number of males, then the former figure is used.

10. Edge clusters (Box 3.12)

In the CBC, clusters that lie partly outside the plot are all counted because the maximum number is needed to obtain the most precise index. If densities are to be described, then dividing the number of clusters by the plot area will produce an exaggerated figure. This is because the number of pairs is actually using an area of land rather larger than the plot. The convention recommended by the IBCC is to include edge clusters if, but only if, more than half their registrations lie within or on the plot boundary. If the boundary is a feature such as a hedge, then all records in the hedge are counted as on the boundary. This method will still tend to exaggerate densities because a species with a territory extending well beyond the plot might still have most of its registrations on the plot. On visits when the bird was elsewhere, it would simply not have been recorded.

There are two other ways to deal with edge clusters. They might on average extend equally within the plot and beyond it. Each territory that includes the edge of the plot is therefore counted as a half. Alternatively, all the edge clusters can be sketched in allowing rings of typical size for their species in cases where a cluster probably extends further beyond the plot than has been recorded. The proportion of the area of each cluster that falls inside the boundary is then counted towards the plot total. Proportions might be estimated by eye to the nearest tenth.

These considerations emphasise the importance of dealing with edge records properly during the field-work. Nearly square or round plots whose boundaries are not particularly rich in birds minimise their number. The most satisfactory approach is to record well beyond the edge of the plot, perhaps by 50–100 m, which helps to interpret clusters that lie across the boundary.

Assumptions

1. The observer is good at finding and identifying birds

This almost goes without saying for any bird census method. In dense vegetation, mapping relies considerably on the detection and recognition of songs. In more open habitats, birds may flee from the observer or skulk quietly, and most of the records may have to be detected, identified and mapped at some range. The main challenge to acuity is in being able to detect and map the more distant records to maximise the numbers of simultaneous registrations. Mapping is not quite as demanding of identification skills as transects or point counts (Chapters 4 and 5) because it is possible to deviate from the route to check anything uncertain. Familiarity with the plot and its birds after a few visits also helps, compared with point counts or transects where the same areas will not be visited repeatedly.

2. Records are plotted accurately

Inaccurate plotting greatly increases the chances that clusters will not be interpreted correctly. Good plotting is helped by a good map with enough features marked on the ground and on the map. In woodland, a compass may be essential. If the distance and thus location of a singing bird is uncertain, it might be possible to move further along the route and try again or to triangulate it with a compass. Familiarity with the geography of the plot helps. Especially in woodland, it is useful to be able to check this out in winter. It is quite surprising how fast the views shorten and change as the leaves come out.

3. The standard rules are used, or broken selectively

The factors known to cause bias such as time of day and year, weather and speed of coverage should be standardised as far as possible. The number of visits should also be close to the standard unless for good reason another number is chosen. It might be tempting to make more visits, but this generally seems to add more confusion than useful further data. The best way to improve the useful records is to give close attention to simultaneous registrations during the field-work.

Interpretation of the maps must also be done carefully. The most arbitrary part of the rules concerns the use of an expected territory size to help in resolving ambiguous data. The best way out of this problem is to try to minimise the collection of ambiguous records in the first place.

4. Birds live in pairs in fixed, discrete and non-overlapping ranges

This is the most critical assumption of the mapping method. Rather little is known about its general realism. Of the birds so far studied, there is quite a range of exceptions. Many non-passerines either violate this assumption or range so widely that mapping plots are too small to encompass more than part of a territory. Species such as doves, corvids, finches and hirundines also have patterns of ranging and territoriality not suited to the method. Poly-territorial species such as Wood Warbler or Pied Flycatcher risk being counted twice for each male. In marshes, the numbers and loose territoriality of species such as Reed and Sedge Warbler make the chances of mapping territories rather remote except in low density areas. However, a good number of songbirds do appear to breed in fixed and non-overlapping territories.

Accuracy of the finished result depends on a high ratio of good records of territory owners to spurious records. The latter might come from wanderers in species where activity is not confined to a restricted and defended range. They might also come from genuine non-breeders. There is increasing evidence that this class of bird is quite common in some species. The mapping method not only fails to count them, it also risks obtaining spurious results because of their presence.

5. There is a reasonable chance of detecting a territory holder

The need for a high ratio of good records to spurious ones also calls for a good chance of finding a territory owner. Clusters with few records must be excluded in case the records are spurious, but real territories with few records may thus fail to qualify as well. The birds that sing clearly meet this assumption best but it should be noted that the season for some is very brief. Nocturnal species are not well counted by mapping because of failure of this assumption unless suitably timed visits are made.

Examples of the use of territory mapping

1. Population monitoring in Britain

The CBC has been run by the BTO since 1962 using the mapping method. It was originated at a period of concern about the impact of habitat changes and agrochemicals on farmland birds. It aims to measure the natural fluctuations in numbers of common birds and to detect any long-term trends. Results have been summarised in Marchant et al. (1990). A subsidiary aim has been to generate data on bird distribution in relation to habitats.

Methods are described in Marchant (1983) and discussed in Marchant et al. (1990). Study plots are categorised as woodland or farmland. About 100 of each are recorded annually by observers who are mainly amateurs. Observers select their own plots and are asked to pick plots representative of farmland or woodland in the region. They are asked to follow rules which are clearly set out and are largely as set out in this chapter. The maps are analysed centrally by a small number of people who are trained for consistency.

Population indices are calculated for species with sufficiently large samples. They are based on the year-to-year changes of numbers of territories summed over all plots that were consistently covered in both years. If observers change or methods deviate from the standard, then the data are not admitted for the years involved. An index for each year is calculated by applying the percentage change from the previous year to the previous index. The index is arbitrarily set at 100 for all birds in a datum year (currently 1980). Long runs of these indices show patterns of population change over time (Box 3.13).

The design of the CBC deals with much of the bias in counting by using paired years from the same observers with standardised methods. It does not matter if a particular observer tends to find more or less than another would, provided this bias is consistent. There is a high level of consistency between observers in measuring year-to-year changes. The mapping method is less susceptible than point counts or transects to the effects of the weather during visits.

The system has weaknesses in that it is not known exactly how results might represent changes in total bird numbers in the country. Nor is it known whether observer choice in starting and stopping work on a particular plot might bias the resulting population changes. If plots with major habitat losses were abandoned, then the indices might not reflect their impact on overall bird numbers. The other disadvantage of the CBC is that it is very demanding of time from volunteer observers and from professional analysts. Point count and transect systems in other countries (Chapters 4 and 5) are based on cheaper methods.

2. The distribution of birds in coppiced woodland in relation to vegetation age

Two studies (Fuller and Moreton 1987; Fuller *et al.* 1989) have related the distribution of birds to the time since coppicing in woods in Kent. A problem with coppicing is that it occurs in small blocks (medians 1.2 ha and 0.3 ha, in the two studies, respectively: range in the first case 0.3–2.7 ha) so that it is difficult to obtain samples of adequate size. The total study areas were 22.3 and 30 ha and had been surveyed for 10 and 5 years, respectively. The mapping method was used so as to associate bird records with coppice plots of known age. It would have been difficult to conduct these studies by any other method because of the small sizes of the study area.

In the first study, bird densities were estimated by attributing each territory to the coppice plot in which it mainly lay, and summing by age classes over all years. It is surprising, however, that it was realistic to attribute individual territories to year classes of coppice—one would not expect the birds to show such close correspondence to a forest plan. In the second study, registrations were counted and summed by age classes over years. This approach avoids the time-consuming problem of drawing and locating territories. Sample sizes were increased by making 23–25 visits per summer thus deviating from the conventional use of the mapping method. A problem with increasing the apparent sample size by making more visits is that many records will be from the same individual birds, so one cannot generalise from the results from the sample plots.

Both studies reached the same general conclusions that species differed in their preferred age classes. There was a general tendency for migrants to be more abundant in the earlier stages and resident species at greater ages. Results for several species are presented in Chapter 10.

Summary and points to consider

Mapped counts are time consuming to complete in the field and to analyse. Their best feature is that, unlike other techniques, they produce a map of distribution of birds.

Is this feature going to be used?

If not, would transects or point counts be more efficient?

Combined with colour-marking and nest-finding, mapping counts have the potential to give good absolute estimates.

There are fixed rules for mapping censuses. Need they be used?

If fixed rules are used, they must be used strictly.

Simultaneous registrations are the key to good mapping.

Can the study be restricted to a limited set of species?

If so, there are variants of method to consider:

Tape-recorded playback

Consecutive flush

Nest-finding

Marked birds

The guidelines for analysing maps need careful consideration.

4

Line Transects

Introduction

The idea of walking about and counting all the birds detected has the appeal of simplicity. One would expect to count more individuals of a species in its favoured habitat than elsewhere and more in a year of high than low population density. By keeping moving, it is possible to cover more ground in a fixed time than by any more elaborate method. Large sample sizes can be generated efficiently. Long transects can be divided into small sections whose habitats can be measured.

This chapter describes the methods used in the field and in calculating relative densities. The critical assumptions are described with hints on how to minimise their violation. Three examples are given of line transects in use. One such study provides widescale monitoring of birds in Finland. One example makes status assessments for seabirds offshore in the North Sea and the third is intended to describe the influence of habitat change on bird communities in North American shrub-steppes.

It is fanciful to suggest that simple transect results are more than indices of relative abundance. There are, however, ways of generating relative density estimates. Various assumptions can be made about detectability of birds which can be used to remove bias of density estimates for comparisons between species or habitats. The assumptions all involve some measuring of distances between bird and observer. As ever, there is a trade-off between the complexity of field methods and the type of results produced: a study needs methods adequate for its purpose.

None of the field methods for transects has been standardised beyond particular national schemes. Indeed, it would probably not be possible or desirable to standardise them because different habitats, bird species and study objectives need different methods. This lack of standardisation, however, has the disadvantage that it is hard to compare results across studies.

Line transects are best suited to large areas that are relatively uniform within sections of hundreds of metres or more. To avoid double counting of birds detectable at long range, transects need to be fairly widely spaced. For these two reasons, the approach is not very good in small areas or for detecting the effects of fine-grained habitat variation. In dense habitats, it is often difficult for an observer to detect birds while moving and point counts may be preferred (see Chapter 5).

The theoretical basis of transect counting can also be applied to detection of signs of birds, such as droppings. It can also apply to transport methods other than foot, with specialist applications in aerial and ship-based surveys.

Detecting and identifying birds while walking is a challenge to ornithological skill. The approach is thus sensitive to bias from observer quality and experience. It is also susceptible to bias from factors affecting detectability of birds which must be controlled where possible and understood in the interpretation of results if control is not possible.

Transect methods can be used year round. Their interpretation in this case needs to consider assumptions about the seasonal variation of detectability due to bird behaviour, weather and vegetation.

Transects are probably more accurate than point counts. This is because the most likely violations of assumptions concern distances between bird and observer. Their impact rises linearly for transects and by square for point counts.

Field methods

1. Routes, visiting rate and travel speed

Routes are selected in accordance with the aims of the study but are usually constrained by accessibility. There is a risk that bias is introduced as a result of selecting for easy access. On farmland, for instance, it is easier to walk the field margins. This will not give a good estimate of overall densities because most species either prefer or avoid hedges. To avoid counting the same birds twice, routes need to be reasonably spaced. In enclosed habitats, this means no closer than about 150–200 m. In open areas, at least 250–500 m would be needed. If the study area can be sufficiently large, it is probably better to avoid having adjacent routes altogether. Another route several kilometres away will produce more thoroughly independent data than one within a few hundred metres of previous coverage.

Routes might be of any total length. In an ideal study, single sections would not adjoin, but all would be separate and independent (see Box 4.1). In practice, spending time and energy in moving from one place to another after completing a single section would rarely be a realistic idea. Routes might be of a fixed length so that each could be covered in a single session of fieldwork.

The total length of route covered depends on the aims of the study and on the resources available. As a rough guide, a minimum of about 40 registrations is needed for a reasonably precise estimate of density of any one species. Clearly the effort needed to estimate densities of any but the most numerous birds is considerable.

For many analytical uses, it might be sensible to divide routes along their length into fixed intervals. Their length would be greater in more uniform or species-poor habitats such as moorland: perhaps up to a kilometre. In richer or more varied habitats, they might be as short as 100 m. If records are separated into such divisions, some estimation of the variability of the results is possible. If habitat features are also measured in the same intervals, then some interpretation might also be applied.

Routes are generally visited once or a small number of times. In a breeding survey, it might be sensible to have two visits to catch the peak activity periods of sedentary and migratory species. Repeating the counts allows some chance of assessing how much more information is being gathered as a result. In general, repeat visits will count many of the same birds. Thus, although the sample sizes appear to be increasing, this is not creating a real increase in precision because the counts are not independent. Rather than count the same individual birds several times, it will often be preferable to use the time to include more different routes.

The speed of walking the routes depends on the numbers of birds present and any difficulty in recording them all. In open habitats, a speed of about 2 km per hour might be reasonable. In thicker areas with greater difficulty in recording all the birds, half this might be reasonable. Speed should be standardised within any one study to avoid adding bias to comparisons between years, sites or whatever.

2. Field recording

The counts used for analysis might be the total (or average) of all registrations or the highest count for any species on any visit per transect section. The reason for the latter is that there are many reasons why a particular visit could produce a low count. In a breeding season survey, it is unlikely that more birds will be counted than generally occur in the area. It does not make much sense, for instance, to take the average count from two visits for a migrant species which had not even arrived on the date of the first visit. Taking a maximum count rather than a total also ensures that spurious precision is not added from an apparently larger sample size which in fact consists of all the same birds counted twice.

Birds can be counted within sections or sketched on a map. Mapping is necessary if there is some uncertainty as to how the next stage of analysis is to be completed. This might include uncertainty over suitable lengths into which to divide the routes because the extent of variation in density is not known in advance. Mapping keeps open the option of changing some of the methods of analysis. It has the disadvantage that the data need a subsequent stage of work which could be avoided if the appropriate totals were recorded on a suitable form in the field. It might also be easier to measure distances from a map than to estimate them in the field. Field estimates have a marked tendency to be rounded to end in a zero or a 5. This can be avoided by deciding in advance to estimate distances to the nearest 5 m.

There are options over what to record by way of activities. Some studies of breeding birds have attempted to get closer to estimates of pairs or singing males by recording sex and activity and treating the records differently. The potential variety of options is limitless. Any such elaboration, however, complicates the field-work and adds assumptions at the analytical stage. Unless such assumptions and the need for invoking them are well founded on

---

Selection of transect routes.

**Box
4.1**

(a) 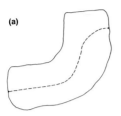   (b)    (c)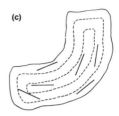

(a) A single route following a natural path is the simplest way to sample an area. The design has the advantage of being the quickest and easiest to follow in the field. It may be the only possible one if access or time is limited. It has the disadvantage that the route may not be typical of the whole area—the path may follow a feature such as a stream or a contour line which affects bird distribution and numbers.

(b) The same length of route divided into six randomly located parts. This design would give a truer representation of the bird fauna of the sampled area if that was the requirement of the study. Since the six routes are independent, it becomes possible to estimate the precision of the resulting mean counts or relative densities for species. The disadvantage of the design is that it would be harder to cover in the field and would take longer because of time spent moving between routes.

(c) Six routes are randomly located with two in each of three classes of distance from the edge of the plot in a study investigating the effects of edge and interior habitats. This stratified random design is often the best if there is some prior knowledge about factors causing variation and those factors can be geographically located in advance.

---

particular knowledge of the species involved, it is probably not a good idea to add such complication.

3. Distance measuring

There are many ways of generating relative density estimates from transect counts. They all depend on some measurement of distance of the bird from the route (see Boxes 4.2 and 4.3). Distances can be estimated within belts (say 0–10 m, 10–20 m, etc.) or each measured individually. In all cases the critical distance is perpendicular from the transect to the bird, not from the observer to the bird. A bird that flushes 200 m ahead, but on the transect route, counts as distance zero. A common method of recording distance is to use two belts, near and far, which can be analysed in one of two ways (see below). The next possible method uses several belts up to about five. Most sophisticated of all is to estimate exact distances (which corresponds to a great many belts). The more belts there are, the harder the estimation of distance becomes in the field and the more elaborate the analytical approach becomes. Using two belts is a good compromise. The width of the inner belt should be such that about half the records fall within it and half beyond. In scrub or woody vegetation this might be at about 50 m, half on each side of the route. In open habitats, a near belt might be more like 200 m across.

Box 4.2   Methods for measuring distance in transect routes.

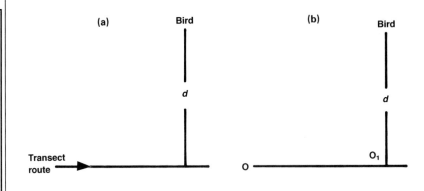

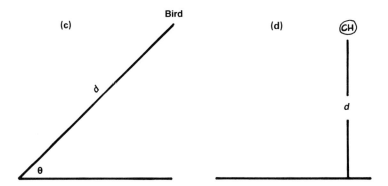

Whether distances are recorded completely or in belts, it is the perpendicular distance between the bird and the transect route which should be measured.

(a) Distance (*d*) is estimated by eye kept in practice with periodic checking against a measured distance. Marker posts might be set out to help.

(b) The observer (O) remembers where the bird was and measures the perpendicular distance (*d*) with either a range-finder or a tape measure when opposite it (O_1).

(c) Distance (*d*) and angle (θ) from the route are measured with a range finder and a compass so that perpendicular distance can be calculated ($d \cos \theta$). This system is not very good for birds close to the route but a long way ahead when detected.

(d) Records are plotted on a map and distances (*d*) measured subsequently. Good mapping is required and may be aided by fixed markers.

Use of bands for recording distance in transect counts.   **Box 4.3**

(a) No distance measuring; all birds are counted. This method is simple but different species are counted on different scales because of differing detectabilities. Five birds ($\times$) have been recorded.

(b1) Fixed belt. All birds are counted within a pre-determined fixed belt (near belt). This lowers the total count but removes distant records of the more conspicuous species. There is little to recommend this approach because it would be optimistic to assume that differences in detectability would not still have a large effect. If the belt was small enough to make high probability of detection realistic for all species, then most sightings would have to be rejected. In this case, four birds have been recorded and three birds have not been recorded.

(b2) Two belts. All birds are counted but attributed to one of two belts. This is an effective method which is very simple to use in the field. Relative densities can be estimated. Four birds have been recorded in the near belt and three in the far belt.

(c) Several belts. Birds are attributed to one of several belts of fixed width (d_1–d_3). This is harder to do in the field because distances have to be estimated to greater precision. It is often more satisfactory to use the methods given above or below. Counts in the first four belts were 1, 3, 2, 1.

(d) Distances are measured to all birds. Distances are perpendicular to the route even if the bird was ahead when detected. This is the hardest method to use in the field but generates the best data for estimation of densities. Birds were recorded at distances d_1 and d_2.

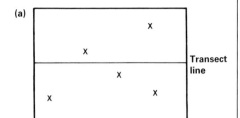

(a)

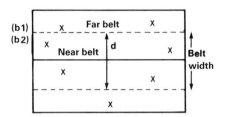

(b1)
(b2)

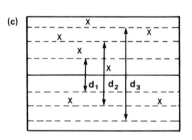

(c)

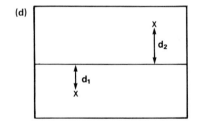

(d)

Box 4.4

Differences in detectability between species.

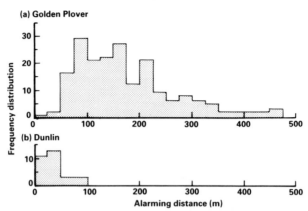

(a) Golden Plover

(b) Dunlin

Numbers of (a) Golden Plover and (b) Dunlin, detected at different distances from the observer on moorlands in northern England are shown (from Yalden and Yalden 1989).

Golden Plovers are noisy and conspicuous, especially as they react to an observer and give alarm at greater distances than do Dunlin.

Dunlin are cryptic and sit tight, so are not detected beyond 100 m. This difference between species needs to be taken into consideration in the design of appropriate breeding water sampling methods.

Box 4.5

Simple ways of dealing with detectability and distance.

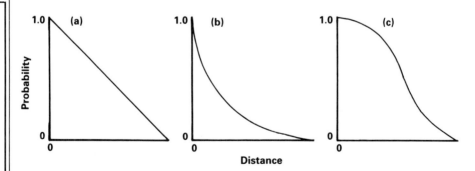

The number of birds detected declines with distance from the observer. This can be expressed as a decline in the probability of detection of a bird which is present. The shape of this relationship is unknown, but three plausible models can be suggested: (a) linear, (b) exponential, (c) half-normal. With data counted in two belts, an actual line can be calculated for any of these assumed general shapes. In practice, all three have been shown to give similar density estimates. The mathematics of the first two are simpler. Note that in all cases, it is assumed that a bird actually on the transect line is always detected (probability = 1).

Calculation of densities from the two-belt method.

Box 4.6

Definitions

Transect length (km)	L
Centre to inner band (m)	w
Total birds	N
Number within w	N_1
Proportion within w	$p = N_1/N$
Density birds per ha	D

Linear model

Probability of detection of a bird at x metres $= 1 - kx$ where k is an unknown constant. Note that at distance $1/k$, the bird will not be detected.

It can be shown that $\qquad\qquad p = kw(2 - kw)$

Hence $\qquad\qquad\qquad k = (1 - SQRT(1 - p))/w$

Density $\qquad\qquad\qquad D = 10Nk/L$

Negative exponential model

Probability of detection of a bird at x metres $= e^{-ax}$ where a is an unknown constant.

It can be shown that $\qquad\qquad p = 1 - e^{-aw}$

Hence $\qquad\qquad\qquad a = (-\log_e(1 - p))/w$

Density $\qquad\qquad\qquad D = 5aN/L$

Example

Transect length $L = 150$ km; inner band $w = 25$ m
Total birds counted $N = 119$; number within 25 m $N_1 = 17$
Hence $p = 0.1428$

For the linear model	$k = 0.00297$
Density	$D = 2.35$ birds per km^2
For the exponential model	$a = 0.00616$
Density	$D = 2.45$ birds per km^2

Note: beware of the units. If the transect length and band widths are in metres, the answer is in birds per m^2, which will be a very small number. It must be multiplied by 10 000 to get birds per ha or by 1 000 000 for birds per km^2.

**Box
4.7**

Densities from full distance recording.

Estimating densities from data where perpendicular distances to all records have been measured is statistically the soundest approach. A full treatment is given by Burnham *et al.* (1980) and a computer program by Laake *et al.* (1979).

The method can be illustrated through a simple example. A transect 1000 m (L) long recorded 40 (n) birds, whose perpendicular distances (x) were measured in metres. Birds beyond 65 m were excluded. (This cut-off distance (w) is arbitrary; it is best if w is set to include 97–99% of the birds detected.)

Values of a_1 and a_2 are calculated from the equation

$a_k = (2/nw)\{\Sigma \cos(\pi kx/w)\}$, for $k = 1, 2, 3$, etc.

(Note that one works in radians for the cosine, not degrees.) The calculations can be laid out in a table:

Bird No.	x	$\pi kx/w$ ($K=1$)	$\cos(\pi kx/10)$ ($K=1$)	$\pi kx/w$ ($K=2$)	$\cos(\pi kx/w)$ ($K=2$)
1	24.10	1.165	0.395	2.330	−0.688
2	32.92	1.591	−0.020	3.182	−0.999
3	8.53	0.412	0.916	0.825	0.679
$\vdots$					
40	7.95	0.384	0.927	0.768	0.719
Total			10.586		−2.888
a_k			0.00814		−0.00222

We then calculate a critical value,

$1/w(2/n + 1)^{1/2} \geqslant \mathrm{abs}(a_{m+1})$. In this case, $0.003398 > \mathrm{abs}(a_2) = 0.00222$.

Since the absolute value of a_2 is less than the critical value, we calculate

$f = a_1 + 1/w = 0.00814 + 1/65 = 0.023$.

If the absolute value of a_2 had been greater than the critical value, we should have calculated a_3, a_4, a_5, etc., until we reached one whose absolute value was less than the critical value. The sum of all the a values, excluding this final one, would then be used in place of a_1, to calculate f.

Density can then be calculated from

$D = nf/2L = (40 \times 0.023)/(2 \times 1000) = 0.00046$ per m^2 = 4.6 per ha.

Confidence limits can also be calculated.

In practice, one should always do such calculations by computer: hand calculation can lead to substantial rounding error.

Population trends of Willow Tit and Capercaillie in Finland in winter measured from transect counts (from Hildén 1987).

Box 4.8

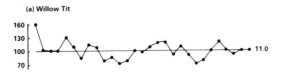

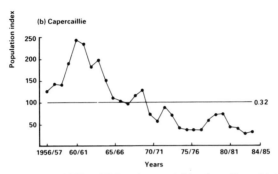

(a) The population of the Willow Tit has been stable since the mid 1950s.
(b) The population of the Capercaillie has been in decline since the early 1960s.

A counting system for seabirds at sea (from Tasker *et al.* 1984).

Box 4.9

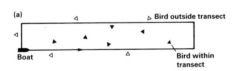

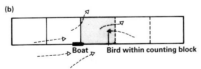

(a) Birds sitting on the sea are counted in blocks of 10 minutes of travelling so that the size of the block is related to the ship's speed which must be known. One side only is counted so the observer has to scan 90° out to a distance of 300 m. In the illustration, six birds are counted and another five seen but not counted.
(b) Flying birds are counted instantaneously in blocks 300 m wide and as far ahead as the observer thinks all are visible. The total number of such scans in each 10-minute period is such that they add up to the same area covered as for the counts of sitting birds. In the illustrated case, there are five birds flying nearby but only one is actually in the box at the instant it is counted. About five such spot-counts will have to be made in every 10-minute period.

Box 4.10　Populations of American shrub-steppe birds affected by alteration of the habitat (from Wiens and Rotenberry 1985).

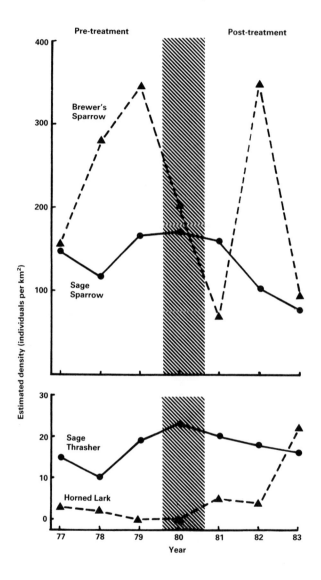

Estimated population densities are shown for four major breeding species in Guano Valley before and after herbicide treatment in 1980 (hatched). Clearly an alteration in the habitat has had an impact on the populations of these species. The herbicide treatment had a negative effect on the population of Brewer's Sparrow immediately after the treatment, but this species rapidly recovered its population level. The Sage Sparrow went into slight decline following the treatment, the Sage Thrasher remained at a similar population level, and the Horned Lark increased somewhat.

4. Special variants

Various methods can be used to increase the detection rate of birds on transects. In open habitats, dogs might be used to flush sitting birds. Rope-dragging can also be used to flush close-sitting birds. This is very hard work and is only worthwhile where nests occur at fairly high densities. In some cases (see Chapters 7 and 8), indirect signs such as wildfowl droppings or seabird nests or burrows might be recorded rather than the birds themselves.

As well as on foot, transects can be conducted from a car, boat or aeroplane. The same general considerations apply to such methods. Counting from a car is particularly good for large and conspicuous birds which occur at low densities, such as raptors.

Counting from a plane is commonly used for waterfowl where access would be difficult or impossible by other means. A plane has an advantage over a boat where either could be used because it is easier to avoid double counting birds that have previously been flushed and have moved a short distance. A plane is also rather more effective at flushing birds, such as some ducks, which might be hard to see in dense vegetation. Air counts are usually conducted with two observers so that one can count each side of the route. Cooperation with the pilot is needed to locate records to routes. The normal system is to fly a prearranged pattern and to time the passage of landmarks, turning points and bird counts. With modern equipment, end points of transects can be programmed into the plane's navigation system. A tape recorder is usually used because counts may accumulate very fast and there is no time to look down and write notes. Flights are normally conducted at a height of 50–100 m and a speed of about 150 km per hour. If the height is kept constant, marks on windows and struts can be used to indicate distances on the ground.

Seabirds may be counted from a ship or boat. The highest possible forward looking vantage point is needed. Some birds are attracted to ships and these may need to be recorded separately. As from the air, it is usual to count by time period and locate the times from knowledge of the timing of the voyage. Most ship-based surveys to date have not used the more sophisticated methods developed for land birds. Seabirds are generally detected because they are moving. This causes considerable difficulties as their flight speed and direction relative to that of the ship will influence the results. Gaston *et al.* (1987) provide some elaborate ways of dealing with this problem. The suggestions of Tasker *et al.* (1984) are simpler (see later).

Briggs *et al.* (1985) compared plane and ship surveys of seabirds. More birds were fully identified from a ship than from a plane, where some had to be grouped. Air counts generally gave higher densities. This was because the air surveys covered a strip 50 m wide while the ship surveys counted birds up to 400 m away (only 150 m for smaller species). Smaller auks can probably avoid detection at such a range even in good sea conditions. The amount of data gathered per unit of time is much higher from a plane. Briggs *et al.* concluded that an aerial survey is more effective for most descriptive purposes. Ships come into their own if other data are to be gathered on bird

behaviour, hydrography or biological sampling. To many researchers, however, the cost of using a plane may be prohibitive.

Interpreting counts with distance estimates

1. Single counts

Single counts may be taken either to infinity or to some pre-ordained
distance. Counting to infinity has the advantage of using all possible bird
records. The disadvantages may be that more distant birds were not in the
same habitat as recorded along the route. Different species are quite obviously counted on very different scales by such an approach (Box 4.4).
Whether or not this is a sensible design will depend on the objects of the
study and the nature of the ground covered.

Single counts within a fixed band give smaller numbers, but they have the
advantage, if this is needed, that the birds were at least within the habitat
described. Although the effect may be less than with counts to infinity,
different species will again be counted on different scales. Although it is
sometimes done, there is rarely justification for dividing the count by the area
(length of route times band width) and calling it a density. This assumes that
all birds are detected and none has fled from or been attracted to the
observer. Such an assumption might be valid if the band is narrow and the
counting unit is the nests of a colonially breeding species (Chapter 8), but
even in such a case, detectability falls off very fast. Thirteen per cent of duck
nests were missed when an observer searched a band of 2.46 m on either side
of a transect and the band would have to be 1.54 m on either side for 100%
efficiency (Burnham and Anderson 1984). For live birds of most species, an
assumption of complete detectability at any distance from the observer could
not generally be warranted.

2. Two belts

If birds are counted in two belts, it is possible to assume the general shape of
the relationship between distance and detectability (see Box 4.5) and use this
to estimate relative density. Three shapes are commonly used. Detectability
may fall linearly from one at the transect line to zero at some distance from it.
Alternatively, a half normal function might be assumed. In this, detectability
falls off slowly at first, rapidly at some distance and then again more slowly.
A third possibility is a negative exponential which has a concave shape.

Estimates of relative density can be obtained with simple algebra using
these assumptions (Box 4.6). They have the effect of adjusting the rank order
of abundance of different species as judged from the total counts. An
inconspicuous species moves higher up the ordered list of abundance. Most
of its records will have been close to the route so a smaller area is being
counted than is the case for a noisier or more visible species.

This method has been used for a long time in Finland (see Järvinen and

Väisänen 1975) and has been widely explored. Järvinen and Väisänen (1983a) provide a table of numbers of birds recorded within and beyond 25 m of the observer. This data set comes from many habitats and observers. These figures could be used for smaller surveys to correct density estimates where the data are insufficient to estimate detectability functions. Different figures are provided from different geographical regions so there is a choice of the most appropriate corrections to use. Before using such an approach, however, it would be wise to check that the Finnish results are appropriate to the new study. This could be done by comparing the near and far counts for the more abundant species with those from the Finnish samples of the same species.

Standard errors of estimates can be estimated from two-belt surveys. A short cut is given in another useful paper (Järvinen and Väisänen 1983b) which shows that the standard error can be predicted remarkably well from the estimated density, the number of transect routes involved and the correction factor (as above). The latter has only a small effect. The standard deviation is approximately proportional to the square root of density and inversely proportional to the number of routes counted. The effect of the correction factor is again small. Järvinen and Väisänen showed that their conclusion was correct in an independent test far to the north in Finland and in different circumstances from the source of the original data. Their equations could thus be used with reasonable confidence elsewhere in Europe. They can also be used in planning future work.

3. Several belts

If birds are counted in several belts, it is possible to plot the counts against distance from the observer (Emlen 1977). They would be expected to decline with distance. The point at which the decline starts can be estimated (by eye or statistically). It will be a different point for each species. Relative densities can then be estimated by dividing the total count within the critical range for each species. The method thus ends up akin to having a single belt count. It differs in that the width of the belt is allowed to vary by species. It will be narrower for inconspicuous species than for noisy or obvious ones. A practical problem with this method is that counts do not always fall off smoothly with distance. Very often the highest count will not be in the central band but at some fairly small distance out from the observer. This is probably because of birds fleeing from the route and not being noticed until they have done so. It is thus sometimes difficult to select the point at which the measured density begins to fall off.

4. Full distance measuring

If distances to all birds are measured, it is possible to make a more detailed model of the shape of the plot of detectability against distance from the observer (Box 4.7). Various functions such as Fourier series, and power or polynomial functions can be fitted. Such methods require computing. For further details of methods and a computer program see Laake *et al.* (1979) and Burnham *et al.* (1980).

A minimum sample size for satisfactory description of detection curves is probably about 100 detections. For many less common species in a community, this requires a very much higher effort than will generally be given in a study. This problem can be overcome by using data from elsewhere for the species or from other species believed to be similar with respect to detectability.

Assumptions

1. Birds exactly on the route are all detected

Even the most sophisticated of distance modelling methods do not work if birds directly on the transect line can be overlooked. This might happen because they are very cryptic in trees above the observer or because they have fled unseen from some distance ahead (thus violating the second assumption as well). If densities are estimated without correction from data from a main belt, then this assumption becomes more likely to be violated. All birds within the main belt must be detected. Because this extra assumption is so unlikely, simple belt methods cannot often be considered likely to measure true densities.

2. Birds do not move before detection

All variants of density estimation assume that birds are randomly distributed with respect to the distance from the route. If they move in response to the observer, then this will cease to be true. Some birds may be attracted, but fleeing is a more likely reaction. In these cases, it is often found that rather more birds occur at some distance from the observer than in the closest region to the route. In the field, particular care should be given to the area immediately ahead from which birds are most likely to flee when they detect an approaching human. In methods using two or a few belts, it is only important that birds do not move from one to another, so the assumption is slightly less restrictive.

Assumptions about movement might also be violated for birds that move a lot in general and are detected only once they come quite close to the census route. Clearly, if the route was walked at an ever slower pace, more and more such birds would be detected merely because they had been given more chance to do so.

3. Distances are measured accurately

This goes without saying but is not easy to do in the field. For birds detected well ahead, it may sometimes be sensible to note a nearby feature and estimate the perpendicular distance from the route once the point is reached.

A tape measure can be used, but this is very time-consuming. An optical range finder is more expensive but very helpful. If distances are estimated, then thorough training and repeated self checking against measured distances should be used. For belt methods, the bird only has to be recorded in the correct belt so the skill demanded in the field is that much less daunting.

4. Individual birds are counted only once

This assumption is hardly surprising but may still cause a problem in the field. If a species is very abundant there is a risk of double counting because of sheer confusion from birds appearing in all directions. Species that move quietly between places where they sing or call might also cause such confusion. The only advice that can be given for field-work is to try to keep track of individuals of the more difficult species. Avoiding problems from this assumption also calls for moving faster rather than slower down the route but this is usually constrained by the need to meet previous assumptions which demand more time.

5. Individual birds are detected independently

The main difficulty with this assumption is likely to arise if birds are more detectable at high rather than low densities. One bird singing or giving an alarm call might, for instance, stimulate others to do so. At lower densities, this is less likely to happen so individuals may be less likely to reveal themselves.

6. Bias from observers, seasons and weather is understood

Transects are more dependent on observer skill than mapping counts where repeat visits give a better chance of finding each pair of birds, and a territory can be registered even if the occupant is overlooked on several visits. Observers can be trained or allocated to routes by systematic design. Seasonal effects might also be large when comparing across years. Routes should be revisited at the same season. Ideally this would be phenologically the same, but in practice calendar date might have to be used if, for instance, timing of breeding is not known. Weather can be handled by applying rules for preventing counts in unsuitable conditions.

Examples of the use of line transects

1. Population monitoring in winter in Finland

Population levels of both wintering and breeding birds have been monitored in Finland by transect methods. The winter project started in 1956 and is the only long-running study of its kind in Europe. Methods and results are described by Hildén (1986, 1987). The aims have been both to monitor population trends and to describe winter numbers and distribution in relation to habitat.

The field methods are very simple. Observers choose their own routes with general guidance. Three counts are conducted each winter within defined 2-week periods. All birds are counted irrespective of the range at which they are detected. About 600 observers take part each year and routes average 11 km in length. Over 10 000 individuals are counted for the most abundant species. Annual counts of the scarce Black Woodpecker have been 91. The simplicity of the method is believed to contribute to its popularity and thus to the amount of data gathered. Results are expressed as individuals per kilometre walked. They can be divided regionally.

Although confidence intervals are not given for individual estimates of this density index, they could be calculated from the data. Inspection of results (Box 4.8) shows that year-to-year changes are often quite small for non-irruptive species such as the Willow Tit. This suggests that the relative densities have been measured quite precisely. Other species show long-term trends or irregular patterns associated with variation in food abundance.

This is a simple study design with methods adequate for its purpose. Open winter habitats and low bird densities lend themselves to transect counts. Because no distance measures are used, the counts cannot be compared across species. Comparisons across habitats would also probably be biased because a greater range would be covered in more open areas. For population monitoring, these problems do not matter so much. There could be a problem, however, due to observers choosing their own routes. Population changes are not computed on a year-to-year basis for the same routes as in the British Common Birds Census (see Chapter 3). If habitat losses were severe, it would be possible for their effects to be concealed by observers picking more interesting areas to count.

2. Distribution of birds in the North Sea

Assessing the distribution and abundance of seabirds at sea has been much in demand especially in relation to recognising places and times when birds are at risk from oil spillages. Birds may be counted by transects from ships. This is not easy because some species are attracted to ships and others flee. Flying birds often violate a critical assumption. If they were all counted fully, a stream of birds crossing the ship's route would lead to an inflated estimate of density. Small or dark species are much harder to detect than the larger and paler species. Those, such as auks, that sit on the sea are harder to detect than those, such as shearwaters, that habitually fly.

Tasker *et al.* (1984) have proposed standardised methods which have been widely used around British coastal waters (Box 4.9). Counts of birds on the sea are conducted in 10-minute periods in a band 300 m wide on one side of the ship and converted to birds per km^2. It is assumed that correction factors for detectability might eventually be devised to convert different species more nearly to absolute densities. Flying birds are counted instantaneously in a series of imaginary boxes. These boxes are 300 m wide and stretch as far ahead as the observer considers suitable for the species and conditions concerned. These counts are repeated at intervals, which depend on the speed of the ship, and can then be converted to densities. If the birds are flying in one general direction, this method can be modified. Counts are made for a minute in a similar imaginary box and the time for an individual to cross it is also estimated. Such counts can then be turned into birds per unit area. Densities of flying and sitting birds can be added. Birds associated with the ship may be recorded but are not added to the estimates of densities.

This approach to counting seabirds may deviate from obtaining absolute densities. To do better would be very much harder and would require more standardisation of viewing conditions than is realistically possible. The essential feature of design is that methods have been standardised as far as is practicable. If everyone followed such a standard, counts of birds at sea would be more widely comparable than is presently the case.

A similar method used in Canada is described in Diamond *et al.* (1986).

3. Habitat of shrub-steppe birds in the USA

Bird communities of the shrub-steppes have been studied using transect methods (Rotenberry and Wiens 1980; Wiens 1985). Individual routes were walked four times in mid June when bird detectability is at its highest. Records were accumulated in several bands up to a range of 244 m on either side of the transect. Densities were estimated by the method of Emlen (1977), discussed above. Habitat data were recorded by laying ten perpendicular transects at fixed intervals across each route and recording at one randomly located point within each 10-m length. The main studies relied on having various different plots surveyed in the same way. The fairly open habitats lent themselves to transect counts. Mapping surveys would have been far more time-consuming.

In one study (Wiens and Rotenberry 1985), the effects of a herbicide treatment and reseeding were reported. The affected plot was surveyed for 3 years before and 3 years after the treatment. Other plots in the area acted as controls. This was not a designed experiment; the authors did not know that their plot was due to be 'improved' but took advantage of the event.

Changes of abundance of four species are shown in Box 4.10. Two analytical approaches were tried. The authors predicted the effects to be expected knowing the vegetation changes that had taken place and having studied the relationship between bird abundance and habitat parameters on other plots. They found that they were not very successful in predicting changes in bird abundance. There were immediate effects on vegetation but bird numbers changed less than would have been expected. Comparisons of before-and-after results at the study site with those elsewhere showed very little consistent pattern from site to site. As a result, it was hard to tell whether changes were due to the vegetation treatment or whether they might have occurred anyway.

There is a clear moral to this story. A short-term before-and-after study on a single plot could be very misleading. If the study had been conducted in the single years before and after treatment, it would have fundamentally mis-identified the effects on Brewer's and Sage Sparrows (Box 4.10). The effects that were expected to occur did, but they did not happen immediately and were further masked by large annual variations. The authors suggest that individual birds may have been site faithful in spite of the habitat having become unsuitable.

Such studies clearly need to be replicated and to be conducted over several years. The authors point out that political and funding considerations often demand instant answers. Those from short-term studies may, however, be dangerously misleading, as this study shows.

Summary and points to consider

Transects are particularly suitable in extensive, open, uniform or species-poor habitats.

They require a high level of identification skills.

They are the most efficient of all general methods in terms of data gathered per unit effort.

Where their use is appropriate, transects can be more accurate than point counts.

Transects generate less detail than mapping counts.

In addition to walking, transect routes can be followed using ships, planes or cars in appropriate circumstances.

Habitat measurements can be made along sections of the routes to coincide with the division of bird records.

There are four variants of measuring distance to birds: one, two, or several belts and complete measuring.

More elaborate measuring is harder to do in the field but permits more elaborate analysis.

A compromise is required to select an appropriate measuring system for the purpose of the study.

There are no fixed rules for transect counts, but careful thought needs to be given to:

   Selection and location of routes
   Number of visits
   Walking speed
   Whether or not to count birds according to activity, age and sex
   Distance estimation
   Observer and other biases

5

Point Counts

Introduction

If you stand at one place, it is possible to count all the birds seen and heard. At its simplest, such a method repeated over several places will assemble a list of species present in an area. With some assumptions about how detectability of birds varies with distance, this can be made into a powerful method of measuring relative abundances rather efficiently (Reynolds *et al.* 1980). The method has come to be widely used for counting songbirds, particularly in France and America. It has some attractive variants if the habitat is also measured in a circle around the census point. Inferences can then be drawn about habitat selection and preferences of individual bird species or communities (Chapter 10).

This chapter describes the methods used in the field and in calculating relative densities. The critical assumptions are described with hints as to how to minimise violation of them. Three examples are given of point counts in use. One such study provides wide-scale monitoring and biogeographical description of North American birds. One example makes a status assessment of a rare bird for conservation application and the third is intended to describe the influence of habitat succession on bird communities.

Point counts are similar in conception and theory to transects (Chapter 4). In fact they can be imagined as transects of zero length conducted at zero speed. They have the advantage over transects of being easier to incorporate into a formally designed study. It is easier to locate points randomly or systematically than it is to lay out transect routes because routes require better access which may bias the habitats sampled. A well spaced sample series of points in an area will provide more representative data than a few transects. Point counts are often preferred to transects in more fine-grained habitats if identification of habitat determinants of bird communities is an objective of the study. This is because the habitat data can more easily be associated with the occurrence of individual birds (Box 5.1).

Point counts are similar to transects in requiring a high level of observer skill. By waiting at each point, there is slightly more time to detect and identify difficult birds than in transects. In some habitats, there is also the advantage of being able to concentrate on birds without the noise and distraction of avoiding obstacles while walking. In scrub or woodland, points may be preferred for these reasons. On the other hand, transects offer a chance to record fleeing birds ahead of the observer. Although they might be seen, such birds cannot be recorded in a point count system if they have disappeared by the time the point is reached and the formal count begins. For this reason, point counts are not commonly used in open habitats or for many larger birds where there are marked problems of fleeing from the observer.

Distance estimation can be applied to point counts in a way analogous to that used for transects. The area surveyed is proportional to the square of the distance from the observer. With transects it is only linearly proportional to lateral distance, with the other dimension coming from transect length. Relative density estimates from point counts are therefore more susceptible to error arising from inaccurate distance estimation or from violation of assumptions about moving birds.

A great advantage of point counts is that they are efficient (Box 5.2). In a single morning one observer might visit 10 points. If the breeding season lasts for about 50 days and points are visited twice then some 200 points can be studied (assuming that the weather is unsuitable on 10 days). In wooded British habitats this would amount to about 2500 records of birds. In the same time it would be possible to conduct mapping censuses on four plots visiting each one ten times. In comparable habitats, this might generate about 500 clusters or territories. So for the same effort, point counts might generate five times more independent bird data for analysis.

A disadvantage of point counts is that there are no standardised methods. On the other hand, standardised methods might be undesirable since different circumstances might be better studied by different designs. Important variables, discussed below, include the number of visits to each point, the measuring of distance to records and the duration of counts. As a result, there are not many published data that can readily be compared from one study to another.

What do densities derived from point counts mean?

There have been very few studies that have tried to compare results from different methods with absolute bird numbers. This is because absolute numbers are extremely difficult or expensive to determine. Many studies have compared different methods, often with the dubious assumption that territory mapping is more accurate than any other.

DeSante (1981) conducted a study in Californian scrubland with complete enumeration by mapping, colour-marking and nest-finding, and compared densities derived from variable circular plots. The point counts underestimated densities but most results were within about 30%. Densities were likely to be overestimated where the species was scarcer and underestimated where it was more common. This is probably because birds in sparser regions probably have larger ranges and make bigger movements round them, so are more likely to be detected. DeSante (1986) has undertaken a similar study in Sierran subalpine forests in America.

Choosing transects or point counts.

Box 5.1

(a) (b)

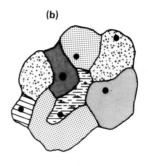

(c)   (d)

X	X	X	X	X
X	X	X	X	X
X	X	X	X	X
X	X	X	X	X

(a) In a fine-grained habitat, such as a wood, a transect following an access route might not be very representative. It would not be easy to divide the bird records into habitat types. Indeed, in this example, two of the habitat types have not been sampled at all.

(b) In the same place, point counts could be set out at random or systematically so as to represent the full range of habitats present in the wood. Each point could also have the habitat recorded around it.

(c) In open country, transects could be set out in a way to cover more of the ground and divided into sections for recording birds and habitats.

(d) The equivalent design for point counts would theoretically record fewer birds but would take about the same time to execute. However, if birds were flushed ahead of the observer, as is generally the case in open ground, this would be a poor design because the observer walking up to a point would scare all the birds away.

Box 5.2

Three different methods for a breeding season census in a hypothetical set of 20 ha oak woods in England.

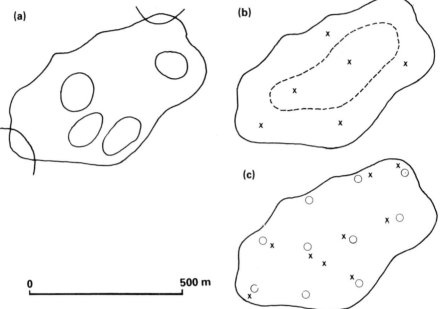

(a) A mapping census. About 150 territories would be mapped for all species. Ten visits would be needed which would take ten mornings. One person could count four such plots in a season. The records would provide the most detailed descriptions of the birds of these sites with most of the species present at least being recorded. Territories would be mapped which would allow comparison with a map of management history or habitat features.

(b) Transect counts. A route of 1 km following existing paths would take about 1 hour to record once. With two visits per route, one observer could record about 40 such routes in the summer assuming they were close enough to allow the coverage of two per morning. In one wood, about 150 bird records (×) might be generated. A longer transect could be fitted into the wood if straight lines were used. This might be difficult to do if the ground vegetation was thick. It could double the number of records in the wood but take so long that it would not be possible to count two such woods in one morning. Habitat records could be taken but it might be difficult to divide the route and the bird records sufficiently finely to reveal the influence of fine-grained variation on birds.

(c) Point counts. Ten point counts (○) would be selected randomly with none allowed to be closer than 150 m from its neighbour. They would take a morning to count including the measurement of habitat features in a 25 m circle round them. Visiting each twice would allow 20 such woods to be counted by one person in a summer. Each would generate about 120 bird records. The description of one site would be quicker but less complete than that from mapping.

If data from many woods are required by the study design, mapping would be too slow. Although collecting slightly less data on birds than transects, point counts might be preferred if habitat associations were also to be investigated, or if there were problems with access, making transects difficult to execute.

Systematic, random and stratified allocation of points.   **Box 5.3**

(a)

A	C	A	B	C	C
x	x	x	x	x	x

B	C	C	A	C	C
x	x	x	x	x	x

(b)

A x	C	A	B x	C	C
		x	x		x, x

B x	C	C	A	C	C
x	x		x	x	

The birds of the rectangular area are best characterised by a random or a systematic allocation of points.

(a) If allocated systematically, three points fall in habitat A, two in B and seven in C which is the most abundant. If allocated randomly, it would also be likely that there would be more points in the common habitat than in the rarer ones but the numbers would probably not be exactly 3, 2 and 7.

(b) A stratified random sample would be better for describing the effects of the three habitats. Equal effort (four points shown) now falls in each habitat. The points were selected by choosing the location of each with random numbers but were constrained so that four fell in each habitat type. In practice, several randomly generated points were rejected after four had been chosen in C but two more were still needed in A. They were also constrained so that no point could lie closer than 50 m from a previously chosen one. Again this was done by rejecting a newly chosen point if it was not suitable for this reason.

Random numbers always look odd but they provide the soundest way of making unbiased statements about the bird communities of the three habitats. Note for instance that by chance one block of habitat C has two points in it while by chance four have none.

Box 5.4

Cumulative percentages of total individual counts with increasing count duration for three Hawaiian species (from Scott and Ramsey 1981).

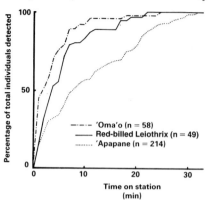

The 'Apapane is the most mobile of the species and the 'Oma'o the least mobile. Half the records of 'Oma'o were achieved in 1 minute while it took 7 minutes to achieve this level for 'Apapane. Longer durations of counts were accumulating records of 'Apapane that had moved during the count which violates a key assumption and leads to an overestimate of density.

Little is known about the details for individual bird species and thus about the practical consequences of such bias.

Box 5.5

Point counts within a fixed radius are not very satisfactory.

(a) Counts to infinity. Although they actually occur at the same density, more of bird A than bird B are detected because A is easier to detect than B. This might be because they sing more loudly or because they are easier to see moving. As a result, there are distant records of A but B is only detected close to the observer.

(b) Counts within a fixed radius. The ratio of abundance of the two species has apparently changed and most records of A have not been used. This could also be because A is also warier than B and has tended to flee from the immediate vicinity of the observer.

Box 5.6

Counts within and beyond a fixed radius.

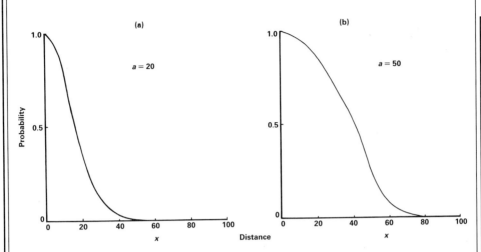

The probability of detecting a bird that is present will decline with distance from the observer according to some curve whose shape is not known. A plausible general equation is: $p = \exp(-(x/a)^2)$ where x is the distance from the observer and a is an unknown constant.

Two plots of this detectability function are shown for $a = 20$ and $a = 50$. Bird (a) is rather cryptic—there is only a 40% chance of detecting it at 20 m from the observer. Bird (b) is much more conspicuous: there is an 85% chance of detecting it at 20 m and it can still be detected at 50 m with a 25% chance.

This curve has the feature of a probability of 1 of the bird being detected at zero range. This is an assumption of the method. Detectability falls off slowly at first with distance and then much faster so that there is not a long tail.

Assuming that the detectability function has this general shape and other assumptions (see text) have not been violated, the value of a can be calculated from the counts for a particular species. This is not actually needed because the density of birds can also be calculated directly thus:

$$\text{Density} = \log_e(n/n_2) \times n/m(\pi r^2)$$

where
n is the total number of birds counted
n_2 is the number beyond the fixed radius (r)
n_1 is the number counted within radius (r) so that $n = n_1 + n_2$
m is the total number of counts
r is the fixed radius

Example: in 326 (m) points there were 421 (n_1) Willow Warblers within 30 m and 925 (n) in all. Their density was therefore 6.09 per ha. Note the units. If r is entered in metres the density is in birds per square metre and needs to be multiplied by 10 000 to turn it to birds per ha.

Box 5.7

Birds are assumed to behave independently of one another. This may not be true.

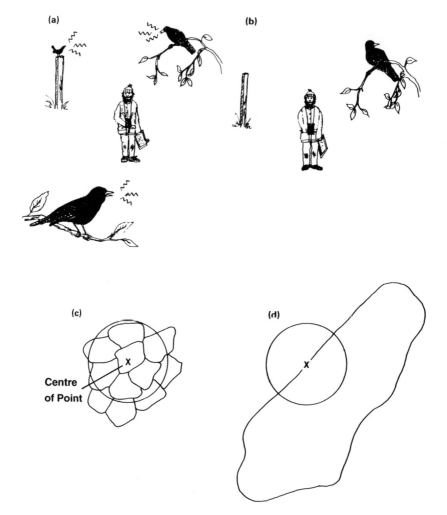

(a) At high densities, individuals may be more vocal as they sing at each other to defend and advertise their territories. Territories are small (c), and individuals may not move far during the count period.

(b) At low densities, individuals may not have a near neighbour and may be quieter. The owner of the larger territory (d) may move further during the count which could lead to an overestimate of density.

Trends in House Finch numbers in the United States from the Breeding Bird Survey (from Robbins *et al.* 1986).

Box 5.8

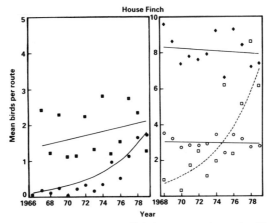

House Finch

The data points are the mean number of birds per census route. The plotted trend lines are calculated in a way too detailed to elaborate here.

Marked increases were noted in the Eastern Region (●) and Southern New England (□) with population levels calculated for Central Region (■), Western Region (◆) and Continental data (○) remaining similar.

Mean relative abundance of the Plain Titmouse and Tufted Titmouse in the United States (from Robbins *et al.* 1986).

Box 5.9

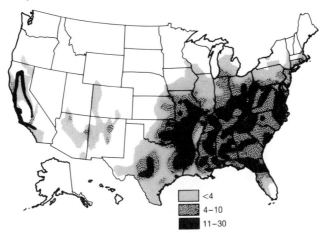

<4
4–10
11–30

In the Breeding Bird Survey of America, the distribution of routes across the continent, and standardised methodology, make it possible to draw general inferences about the geographical variation of abundances of species. This is not hampered by the abundance figures having no absolute meaning. Note the tendency which is shown by many species to be more abundant at the centre of their ranges than at the margins.

Habitat selection by the Chiffchaff in conifer plantations in Wales (UK) (from Bibby *et al.* 1985).

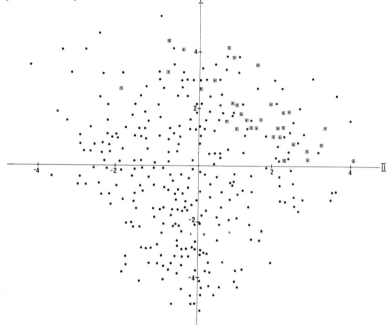

A total of 326 point counts were conducted and habitats measured at the same places. This allowed a variety of analytical approaches to describing habitat selection of the bird community.

In this figure, all the census points are plotted according to their position in a two-dimensional view of habitat. The axes (derived from principal components analysis) approximately coincide to age (Axis I) and conifer/broadleaf mixture (Axis II). Chiffchaffs (marked with squares) occur only in forest stands with comparatively mature vegetation and a high broadleaf content.

Field methods

1. Selection of points

Points to be counted can be laid out systematically or selected randomly within the study area (Box 5.3). Random selection based on a grid of random numbers generated from a computer or a table allows general inferences to be drawn about the area sampled. The sample design may be stratified. One might, for instance, want to study the effect of succession on the bird community in a certain kind of woodland. An appropriate design might be to allocate 20 points randomly in each of five groups of stands representing different stages of succession. Such a stratified approach would be more efficient for such a purpose than simply allocating points randomly in a large area which might under-represent some of the scarcer habitat features of interest.

Points should ideally be no closer than about 200 m apart in woody vegetation. If they were any closer, some individual birds would be counted at more than one point. This would give a spurious inflation of sample sizes and apparent precision of results. If habitat selection is the object of the study, it will be found that the scarcer birds in the community will only occur at rather few points. Habitat selection by such species can be studied by collecting habitat data in a circle around the bird once it has been detected (see Chapter 10). Such data cannot, of course, be used to make inferences about the density of the species involved (which could be derived from randomly allocated points). They can, however, be compared with the habitat data at points where other species have been located. In other words, habitat data are compared at points with and without a particular species.

Because of the need to avoid duplication from nearby points, it is difficult to fit many points into a small area. For this reason, point counts do not lend themselves to describing the birds, or measuring year-to-year changes, in a small area. In a 20 ha wood, all the songbirds might be counted in a mapping survey, but there might be room for only five to ten point counts depending on the minimum separation used and the shape of the wood. The point counts could easily miss several of the species represented by only one territory and would only just be adequate to describe the densities of the commoner species. If the 20 ha wood was a nature reserve and was the object of study, a mapping survey might be chosen. If, on the other hand, the study was about habitat factors affecting woodland birds in a large area, point counts might be suitable. Several woods would be sampled and habitat data collected at the study points (Chapter 10).

In small patches of habitat, thought needs to be given to the inclusion of points near the edges. These will include birds living in adjoining habitats, so open country birds might appear in a woodland bird census. Depending on the purpose of the study, this may or may not be desirable.

2. Duration of counts

Counts can begin as soon as the observer reaches the point or can be delayed for a few minutes to allow birds to settle down from any disturbance caused by the observer's arrival. A possible design, if habitat features are being described or measured, is to do some of this during a settling down period and then start the bird counts. If there is a settling down period, it would not be a good idea to use this for any habitat measuring that involved the observer walking around the area.

Point count studies have used durations as short as about 2 minutes and up to as long as 20 minutes. The longer one stays at a point, the more birds are detected (Box 5.4). Normally, however, the majority of records are accumulated quickly and fewer and fewer are detected in each successive time interval. In a long count, it becomes ever harder to be certain that a 'new' detection is not in fact a bird that was seen some minutes previously and has moved somewhat. Longer durations are also more likely to record birds making long movements which invalidates a critical assumption of the method (see later).

In most temperate situations, a shorter count duration of 5 or perhaps 10 minutes is to be preferred (Fuller and Langslow 1984). The French IPA (Index d'Abondance Ponctuel) uses 20 minutes. This is probably longer than is ideally efficient. Longer intervals might be needed, however, in places with a richer bird fauna or with more species that are very hard to detect, such as tropical forests. Time saved by shorter counts can be used to gather data from more points.

3. The recording method

The objective in point counts is to count each individual bird once and once only. It is, however, possible to separate birds into different categories of means of detection. Some observers have counted individuals by sex and combined the results into pairs or territories. In Finland, for instance, pairs are the counting units. Pairs can be based on a single male or female, a true pair, a flock of fledglings or a nest. If several individuals are encountered the number of pairs is obtained as half the total, rounded up for an odd number. Alternatively it is possible to accumulate all records together. In most circumstances in woody vegetation, more birds are detected by sound than by sight. In the breeding season, many will be singing, but at other times of the year this will not be the case. For most purposes, it seems rather arbitrary to establish elaborate rules for combining records of different kinds. The rules have to differ by species according to whether the sexes can be recognised and whether or not song is frequent.

4. Distance estimates

With certain assumptions (see below) estimates of density can be made if a distance measure is associated with each bird detection (Scott *et al.* 1981). It is always worth doing this and the simplest methods are not too difficult in the field. The most sophisticated approaches are more difficult to execute in the field and involve elaborate computation. There is a trade-off between time and difficulty in the field and quality of the results.

The simplest bird counts with no estimation of distance produce results biased in favour of conspicuous species. Imagine two equally abundant birds. Bird A has a very loud and carrying song—one of the thrushes would be a good example. Bird B has a very quiet song, such as a Goldcrest or Treecreeper. It is obvious that more of bird A will be detected than of bird B even though they are actually of equal abundance. This is because bird A can readily be detected at up to 100 m even in a closed wood, but rather few individuals of bird B will be detected at more than about 30–40 m.

The most sophisticated method of point counting involves measuring the distance from the observer to each registration. If the bird moves, the distance to measure is to the point where it was first detected. In practice, distance measuring is extremely difficult in closed vegetation. It is hard enough to estimate distance from oneself to a fixed visible point. If the bird has been detected by call or song, it is still harder to estimate its distance.

There are compromises between these two extremes. Records can be allocated to one of two or more circular bands of distance from the observer. If two bands are used, a sensible dividing point is at about 25–30 m. If many bands are used, they may be of about 10–20 m in width.

Interpreting counts with distance estimates

The considerations that apply to transects are the same for points, with the exception that errors in assumptions or measurements are more serious. The algebra of the calculations is slightly different.

1. Single counts

Single counts either to infinity or within an arbitrary range, such as 25 m from the observer, provide no more than an index of relative abundance. Different species are measured on different scales, and abundances across species cannot be compared. Large and conspicuous species will be over counted relative to quiet or cryptic ones (Box 5.5). The counts are, however, quick and simple to make in the field. If a fixed radius is used, this bias might be less severe, but the actual area counted is very small. With a counting radius of 25 m, ten points counted in a field session cover only 2 ha. By comparison, an 8 km transect with a similar distance limit would cover 40 ha. The other problem with fixed radius point counts is that many birds might have moved and be just beyond the zone included. Many potential records therefore have to be ignored. This is less difficult with transects because there is a chance of recording birds close to the line as they flee ahead of you.

2. Two counting bands

With two counting bands ($0–r$ and r to infinity) relative densities can be corrected for variation in detectability of species. An assumption is needed about the form of the relationship between distance and detection probability. One plausible assumption and the required calculations are shown in Box 5.6.

3. Several counting bands

The density of registrations should fall with distance from the observer as birds become harder to detect at greater distances. Note that it must be density rather than number because bands of fixed width have greater area at greater distance. The distance (R) at which the decline becomes marked can be identified. Relative density is calculated by dividing the number of records within R by the area of the circle of radius R. In practice, this procedure is not always simple with point counts because the peak density of records is often not in the central circle.

4. Full distance measuring

If the distances to all records are measured, then the shape of the detection curve for each species can be estimated and relative densities can be calculated. The statistics of this approach are quite elaborate and require a computer to calculate. Details are given by Buckland (1987) and are not repeated here.

Assumptions

As with all methods, it is important to understand the assumptions that are made. In this way, steps can be taken to ensure that they are met in the field or that unwarranted conclusions are not drawn if the assumptions are violated. Point counts require the following assumptions.

1. Birds do not approach the observer or flee

This assumption is most conspicuously untrue in open country where very few birds will remain within 10–20 m of an observer. It is also particularly violated by larger species which in general are warier of humans. If a distance measuring approach is being used, the critical assumption is that fleeing birds do not move from one band to another. In practice if the number of registrations per unit area is plotted against distance from an observer it is often found that most birds are some tens of metres away with very few remaining close to the observer. This suggests that the closest recording band used in analysis should be big enough to embrace the abundance of birds that have fled a short distance. A larger band-width is generally required in more open habitats. Violation of the assumption about fleeing will generally lead to underestimates of density. If birds approach the observer, densities might be overestimated.

2. Birds are 100% detectable at the observer

Various assumptions can be made about the rate at which detectability changes with distance but all methods assume that the bird is fully detectable at the observer's location. In practice this assumption is most likely to be violated for very quiet and skulking species which includes most nocturnal birds such as owls, which are barely detected by point counts (or by any other general census method). It is likely that in high forests birds directly overhead could be missed if there is dense foliage between them and the observer. Violation of this assumption leads to an underestimate of density and makes some birds uncountable by this method.

3. Birds do not move much during the count period

If birds were highly mobile, it would be possible to count large numbers of individuals of one species whilst standing at a point. Mobility of birds is possibly one of the reasons why the number of birds counted increases the longer one stays at a point. It also causes considerable problems in recognizing individuals. Imagine a bird calls from one direction, goes quiet for a period and moves, and then calls again somewhere else. It is difficult to avoid counting that individual twice. The problems of mobility are best dealt with by taking a short census period. If the period is too short, however, one may miss the more silent or skulking species nearby. Violation of mobility assumptions make it difficult to compare species such as raptors, pigeons or corvids on the same scale as measures given of songbirds. Densities of mobile species will be overestimated. Point counts can be used in winter but birds are generally then more mobile than in the breeding season so densities might be overestimated. Species moving in parties in tropical forests cause similar difficulties.

4. Birds behave independently of one another

Sometimes one bird reveals itself by call or a song as a result of another individual calling or singing (Box 5.7). The effect of such behaviour might be to make high densities more easy to measure more accurately than lower ones. There may also be an interaction between density and mobility of birds. At lower densities, individuals may have larger ranges and move more, thus violating the assumption above. In other words, it would be possible for the number of birds detected to have a non-linear relationship with the number actually present. There is no known way to deal with the consequences of violation of this assumption which are anyway poorly understood.

5. Violations of the above assumptions do not interact with habitat or elements of study design

It is quite possible that fleeing, detectability or moving behaviour of birds might vary by time of day or habitat. In this case bias might fall differently in different circumstances. Problems with matters such as time of day or weather are dealt with by standardising the method to a fixed range of conditions and ensuring that the analysis compares like method with like. Interactions with habitat are potentially more serious since the effects of habitat variation may well be an objective of the study. It is fairly evident that fleeing behaviour and detectability may vary with openness of vegetation. If no distance measuring is done then counts in different habitats will quite clearly be unrelated to densities in any simple way. If counts have been made with one of the methods that includes distance estimating then it is possible to correct for different detectabilities in different kinds of vegetation.

6. Distance estimates are accurate

Accurate distance measuring is particularly important for point counts because any errors are squared in density estimates. Observers should be trained to assess distances. Training has been shown to have a marked improving effect. If permanent points are counted repeatedly, it is possible to mark some fixed distances to use as reference points. Some studies have used optical range measuring equipment. This is quite helpful for a visible bird but not very good for one that cannot be seen. Some studies have recommended locating calling birds by the observer moving. This has the disadvantage of creating further disturbance and exaggerating the problems of birds fleeing.

If counts are conducted in bands rather than by full distance measuring methods, it is only necessary that individuals are put in the correct bands. This is obviously not as difficult to do in the field as complete distance estimating. In the case of the two-band method, it is only necessary for the observer to have a very good idea of what the single radius looks like and to check each bird as to whether it is within or beyond that distance.

7. Birds are fully and correctly identified

Because most detections are by sound, point counts require a particularly high level of field skill. The option of moving to see and identify a bird is often not available. Observers, therefore, need to be fully familiar with all the species in the area and with the separation of any that sound similar. Training for consistency in identification is almost certainly worthwhile in an area unfamiliar to the observer. Tapes might be used for training and for assessment of standards of observers. If many observers are used in one study, thought should be given to how they are allocated to particular points. In this way, study design could, for instance, prevent bias arising in the counts from one habitat type because one aberrant observer did all the recording there. If points are visited more than once, observers could be swapped. In this way, it would be possible to check their degree of consistency. Point counts are not a very good method for people with hearing loss.

Examples of the use of point counts

1. Monitoring breeding birds in the USA

The American Breeding Bird Survey (BBS) is a large programme which has run throughout the USA and Canada since 1965. It is sponsored by the US Fish and Wildlife Service and the Canadian Wildlife Service but draws much of its support from an organised network of amateurs. The BBS aims to monitor population trends of a wide range of breeding birds (about 230 species). The information generated can contribute to a variety of objectives of differing scales.

(1) Measure normal year-to-year fluctuations in numbers of individual species.

(2) Detect the effects of variations in weather including catastrophic events.

(3) Measure long-term trends in numbers.

(4) Allow description and analysis of fluctuations and trends on a geographical or habitat basis.

(5) Describe widespread biogeographical phenomena such as regional and habitat based differences in relative densities of particular species.

Data are gathered by point counting. Points are located along roadside routes each of which consists of 50 points spaced at about 800 m apart (actually 0.5 miles). They are visited once each summer at a date mainly in June but with some allowance for latitude. The counting starts half an hour before sunrise and each point is counted for 3 minutes. The total number of each species is recorded up to a distance of 0.25 miles away (about 400 m). The results are recorded on pre-printed forms which give a species list and columns for each count stop. Observers also add the records up on a summary sheet which saves administration costs, but the field sheets are also archived.

From year to year, the routes are always run in the same direction. Stop points for counts are mapped and described once established so that they are constant from year to year rather than reliant on a car's odometer (mileometer in UK). No methods of attracting or provoking calls from birds are allowed. The instructions make it very clear that the counts are not expected to be a complete record of all species actually present. Standardisation rather than completeness is essential. Observers are cautioned not to stay longer or otherwise bend the rules in the hope of a 'good' species which they know or suspect should be present.

Forms are returned by the end of July, checked by an editor at the Fish and Wildlife Service, entered on tape and run through a further checking program. Observers receive a printout of their data to check that the official computer entry is correct. Machine-generated summary data are produced annually for a variety of purposes.

The BBS has about 2000 routes which are counted each year generating about 1 500 000 records of around 500 species annually. A remarkable feature of the BBS is that routes are formally randomly selected. Each lies within a single one degree block of latitude and longitude and within one state. The majority lie within a single physiographic unit derived from a map of the life zones of America. Organisers try to find observers to count allocated routes in contrast to the system in Britain where observers select and drop their mapping plots as they please (Chapter 3). Coverage varies with human population density but this can be allowed for in a stratified design. The advantage of such a system is that conclusions can be generalised with confidence by weighting the results from different routes. Other countries have doubted whether a largely amateur effort can be so effectively directed in the interests of good statistical design.

Long-term trends were originally calculated on the basis of ratios of counts for a species on the same routes in two adjacent years. For a variety of reasons, this method has now been abandoned in favour of a statistically more rigorous one. Trends are estimated for each route by a log-linear regression model. In this case, all the data are used for each route but the method can cope with occasional missing values. National trends are compiled as a weighted mean of all the site trends. The weighting factor has three elements to ensure that each region is weighted by area irrespective of number of routes involved, that routes with more birds are more highly weighted and that routes with more data points and thus a better trend estimate are more highly weighted. An example is shown in Box 5.8. Fuller statistical details including description of the estimation of variances is given in Geissler and Noon (1981).

All the count data originating from the BBS are on a relative rather than an absolute scale. Different species are counted on different scales so it is not possible to say that one is more numerous than another by any known amount. Within species it is, however, valid to draw strong inferences about population fluctuations and trends and about patterns of relative abundance in relation to broad habitat type and geographical location (Box 5.9). The design of the study means that such conclusions can be generalised to the whole area and confidence limits can be set. This is not a perfect bird census (such a grail may not even exist!) but one can only admire the range of information it has generated. Further reading can be found in Engstrom and James (1984) and Robbins *et al.* (1986).

2. Estimating the range and abundance of the Azores Bullfinch

The Azores (or San Miguel) Bullfinch is a very rare bird which went unseen for 40 years, was recently rediscovered, and is now known from a very small area of one island in the Azores. The terrain is steep, susceptible to land-slip and covered with impenetrable vegetation. The objective of a 20-day study was to assess the Bullfinch's range and numbers so as to be able to make a start on a conservation plan. Lack of time, lack of prior knowledge and the complete inaccessibility of so much ground made much formality of method difficult.

We chose to use point counts. A mapping census would have been completely impossible because of access problems. We would not anyway have known where to locate plots because of lack of prior knowledge. Transects were considered but access was again a problem. We expected the bird to be very quiet like the European Bullfinch and thought that even the noise of walking over difficult ground might lower the chances of detection. The points were spread at intervals of 200 m along any access routes found in or near the suspected range of the species. Counts lasted for 10 minutes since we expected the bird to be quiet and inconspicuous and possibly able to remain close to an observer, but undetected, for several minutes. This was checked by separating records into two 5-minute periods. Records were allocated to within or beyond 30 m at first detection. Descriptive records of the vegetation were made at each point so that we could compare points with and without Bullfinches.

The method was described in terms of numbers of points covered and their locations, the time of counts (10 minutes) and the radius separating near and far records (30 m). It will be possible for anyone to repeat this survey at any time and tell whether this very rare and localised bird is declining or increasing in numbers.

The results are published elsewhere (Bibby and Charlton 1991). We were able to narrow down the likely range of the bird because we had formally located both where we found it and where we had looked with similar effort but failed to find it. We were able to make some general observations on its habitat for the same reason; we had described the habitat where we found the bird and where we did not. The counts could be turned into relative density estimates by assuming an exponential relationship between detectability and distance. By applying such an analysis to areas in which we found the bird, we could make a first estimate of densities and total area of the likely range and thus of total numbers. The total population estimate of about 100 pairs should be treated as little better than an order of magnitude estimate. The areas that were accessible were not the same as those that were not so we cannot be confident that our sample was representative of bird density across its very small range (about 500 ha). Because counting methods were standardised, we can, however, claim that changes of numbers will be detectable when the same method is repeated in the future.

This was not a perfectly designed study because access did not allow it. This case illustrates that some consideration of methods is none the less worthwhile. Many of the world's 1029 endangered birds (Collar and Andrew 1988) have never been counted, which is a severe obstacle to their conservation (Green and Hirons 1988). Even partly designed and formalised methods are more useful than casual records in the process of assessing changes of numbers and threats.

3. Studying the effects of vegetation in young forestry plantations

The aim of this study (Bibby *et al.* 1985) was to investigate the factors influencing bird occurrence in young forestry stands at the start of the second rotation. There was one summer in which to do the work with two observers. It was important to obtain results of wide generality so all the forests in North Wales were included. Stock maps showed all possible plots below 10 years of age. The only way to collect enough data was to use point counts. Size of area cut was potentially important so study plots were chosen (on a random basis) to cover the range of sizes within the six main forest blocks. The number of points per plot was set to be proportional to size of plot, and individual locations were selected randomly with the constraint that two sample points had to be at least 60 m apart.

Each point was visited twice at an interval of about 30 days because many of the birds were migrants with later breeding seasons than the residents. The two observers were trained in distance estimation and checked that they were familiar with all the likely birds. Further, to remove any observer bias, each observer made just one of the two visits to each point. Vegetation measurements were made at each of the sampled points. Many of the study plots were densely vegetated and difficult of access but 326 points were visited in a 2-month field season. Over 3200 records were obtained of 31 bird species.

The study would have been very difficult to execute by any other counting method. An advantage of point counts is that one can compare vegetation features at points with or without particular birds (Box 5.10). A problem with the method is that estimates of relative densities cannot readily be compared with much other published data. Mapping results have been published more extensively to date.

Summary and points to consider

Point counts are suitable for conspicuous birds in woody or scrubby habitats. They are suitable for study of extensive areas but do not provide the level of detail of mapped counts.

They are more efficient in terms of data collected per unit effort than mapped censuses but less so than transects.

They may be more appropriate than transects in areas where access is poor or where habitat is very fine-grained.

In open country or large-scale habitats with good access, transects may be more suitable.

Point counts have special value in habitat studies when habitat is measured at the counting points.

There are no fixed rules for point counts but careful thought needs to be given to:

    Selection and location of points

    Number of visits

    Duration of counts

    Measuring of distance to records

    Observer bias

6

Catching and Marking

Introduction

Individual birds may be caught and marked in order to estimate population size, to investigate habitat selection and other distributions, to calculate survival rates, measure dispersal and other movements, and to measure the reproductive success of individual birds. This chapter concentrates on the use of catching and marking to aid the counting of birds and the estimation of population size. The methods can be divided into capture–recapture (in which birds are caught, marked, released, and a proportion recaptured), and capture per unit effort (in which the effort used to catch them is standardised, or the capture rate for a species relative to the total number of birds captured is used for calculating population indices). Some developments of studies involving capture, such as the use of matrix models, are also given.

There are many difficulties and assumptions involved in catching and marking birds. These include the need for training, legal licensing, expertise in applying marks, and experimental design. In respect of the latter, for instance, habitat structure, particularly vegetation height, will influence what you catch. However, properly thought out and organised marking experiments can provide information of immense value which would not be produced by any other method.

This chapter gives some guidance as to when it is necessary to mark birds and how to go about it. References to licences and permits refer to work undertaken in Britain. Similar, but by no means identical, considerations, apply to almost every other country. For many mobile species colour-marking systems are controlled through international agreements. These are set up to protect the research workers from duplicating marks and so flawing each other's work and also to facilitate the exchange of sightings. In some countries, including Britain, cooperation with such protocols is mandatory for anyone obtaining a licence to mark birds, and every research biologist considering such marking should cooperate.

Major considerations in catching and marking

A number of assumptions need to be met. Not all methods require that individuals should be recaught. In some cases field observation is adequate.

(1) Will it be possible to catch enough individuals to obtain worthwhile results? Some species are far easier to catch than others.

(2) Will the mark harm the bird or affect its behaviour? Will it make the bird more vulnerable to predation, alter its place in any hierarchy (e.g. feeding), or interfere with pair bonding?

(3) If the bird has to be handled again for the ring to be read can it be caught again? Are its chances of capture affected by the fact that it was caught in the first place?

(4) If the bird is to be observed in the field are the marks used properly distinguishable? They may fade or fall off, your co-workers (or you yourself) may be colour-blind. Is your recording system as foolproof as possible? Is the range at which you can distinguish the marks distant enough to be useful?

Marking methods

These are listed with description, constraints, advantages and disadvantages. Readers are referred to the BTO Ringers' Manual (BTO 1984) for details of field etiquette, catching methods and recording. This information has international relevance.

1. Metal ring

Each split metal ring has a return address and unique serial number. The bird is uniquely marked but must be recaught to establish its identity.

2. Colour ring

Celluloid springy rings are used on small birds but more modern materials (mostly darvic) are used for bigger birds. The most usual style is a spiral-flat band. Rings may be placed above the knee on waders to provide a greater number of colour combinations. Celluloid rings (only material available for small birds below about 30 g) may fade. Larger celluloids rings are cemented or sealed but smaller ones are not.

3. Leg flag

Coloured sticky tape around the metal ring can be used, though it falls off after a few weeks or months. This is a problem since the exact life of any tag applied cannot be known. On most types the flag sticks out about as far as the height of the ring.

4. Patagial tag

A small plastic flap is pinned on the top surface of the wing by a stainless steel pin through the bird's patagium and held in place with a nylon washer. Such tags may be distinguished by their colour, letters, numbers or symbols stuck or drawn on their surface. For many large species such as dabbling ducks they have been shown to be safe but are not recommended for diving birds and have never been permitted on anything smaller than a Starling in Britain. They may not be easily visible as many birds preen them into the wing coverts. Particularly bright colours used on patagial tags may make the marked birds significantly more vulnerable to avian predation, though evidence is sparse.

5. Neck collar

A large colour ring can be put around the neck of the bird. These have been supposedly successfully used on geese and swans but have also been the subject of significant problems e.g. affecting pair bonding or causing physical distress and choking.

6. Plumage dyes

Feathers do not dye easily. The best dye is picric acid which dyes light-coloured feathers orange–yellow. The marks are temporary (until the next moult at the latest).

7. Radio-tracking

This is an unrivalled technique for locating individual birds to determine home-range, time budgets, habitat selection, etc. (see Chapter 9). This equipment is commercially available, and further details of considerations in a radio-tracking study are given later in this chapter. It has the disadvantage that only a few individuals can be tracked at a time by one observer, unless expensive automated tracking is used.

8. Individual marks

Some species of birds have individually identifiable marks which makes recognition possible without the aid of an artificial marker. The bill patterns of Bewick's Swans are a classic example (Box 6.1).

9. Others

Bill plaques, back tags, imped feathers, numbers painted on the faces of swans or the shields of Coot have all been used. These rather specialist and individual marking methods are not described further in this book.

Capture–recapture methods

The two main reasons for catching and marking individuals within a population are (1) to estimate population size and/or migration routes and (2) to estimate survival rate. Analytical methods for mark–recapture data for each of these have been developed and the underlying principles, largely concentrating on population estimation which is the focus of this book, are outlined here.

Estimation of population size is based on the assumption that, if a proportion of the the population is marked in some way, when it returns to the original population complete mixing occurs. A second sample is then taken. The number of marked individuals in the second sample should then have the same ratio to the total numbers in the second sample as the total number of marked individuals originally released has to the total population. As the number originally caught, the number marked and the number of marked individuals in the second sample are known from catching and marking, an estimate of the size of the total population can easily be calculated. Mark–recapture population estimation is particularly useful where absolute estimates of population size are needed for species that are difficult to count in the field.

Ornithologists have not made much use of mark–recapture methods (e.g. Jolly 1965) for estimating population size although reviews can be found in Cormack (1968, 1979), Seber (1973) and Nichols *et al.* (1981). Mark–recapture models can be classified according to assumptions about whether the population is closed, i.e. the population is not influenced by mortality, recruitment or migration (both immigration and emigration). 'Models' or mathematical expressions are used to define two broad classes—for closed and open populations. Such models as the Lincoln index and Jolly–Seber are used to analyse capture–recapture data and this section concentrates on their applicability and assumptions rather than mathematical detail. References to further details of these models are given later in this chapter.

There are four classes of mark–recapture models:
(1) Closed populations:
(a) two-sample experiment (e.g. Lincoln index type model with two capture events)
(b) *K*-sample model (many capture events)
(2) Open populations:
(c) completely open populations (both losses and gains, e.g. Jolly-Seber model)
(d) partially open populations (most commonly losses and no gains, but also gains and no losses)

To satisfy the requirements of the various methods of analysing mark–recapture data for population estimation, and having defined why it is to be performed, a number of assumptions of the methods should be considered. The identifying number of each assumption is related in Box 6.2 to the four types of model described above. The ways in which these assumptions are likely to be violated and possible ways of reducing the problems are presented below.

Naturally 'marked' birds.

Box 6.1

'Yellowneb'
Bewick's

'Pennyface'
Bewick's

'Darky'
Bewick's

Sir Peter Scott, of the Wildfowl and Wetlands Trust, showed that bill patterns on Bewick's Swans can be identified by variations in their black and yellow bill markings, which usually fall within the three major categories illustrated (from Scott 1981). Reproduced with kind permission of Lady P. Scott.

Box 6.2

Assumptions of the four classes (a–d) of mark–recapture models.

For example, models a and b both assume that the population under investigation is 'closed', i.e. there are no additions or subtractions (immigration/births or emigrations/deaths), and so on.

Assumption	a	b	c	d
(1) Closed population	×	×		
(2) Equal capture probability	×	×		
(3) Marking has no effect on catchability	×	×		
(4) Second (subsequent) sample(s) are random	×			
(5) Marks are permanent	×		×	×
(6) All those marked that occur in second sample are reported	×	×	×	×
(7) Capture probability assumed constant for all time periods		×		
(8) Same probability of being recaught in all capture events			×	×
(9) Equal probability of survival			×	×
(10) Equal probability of caught birds being returned			×	×
(11) Sampling time is negligible			×	×
(12) Losses from emigration and death are permanent			×	
(13) Population closed to recruitment only				×

Note: a = Two sample model—closed populations.
    b = *K*-sample model—closed populations.
    c = Completely open populations.
    d = Partially open populations.

| Calculation of the simple Lincoln index. | **Box 6.3** |

The simple Lincoln index is used here to analyse a 'two-sample' case of capture–recapture data to provide an estimate of population size, P. The calculation of standard errors is also shown. The general format is: Number in population/Original number marked = Number in second sample/Number recaptured, or

$$P = \frac{a\,n}{r}$$

where $n =$ number of individuals in the second sample, $a =$ number marked and $r =$ number recaptured.

If the second sample (n) consists of a series of sub-samples and a large proportion of the population have been marked, recovery ratios ($= r/n$) can be used to calculate the standard error (from the variance) of the estimated population where:

$$P = \frac{a}{R_T}$$

where $R_T =$ the recovery ratio (r/n) based on the number of birds in all of the samples; and the variance is approximately:

$$\text{var } \hat{P} = \left(\frac{a}{R_{T^2}}\right)^2 \times R_T \frac{(1 - R_T)}{y}$$

where $y =$ number of birds in the sub-samples. The standard error of the estimate is the square root of the variance.

The method is based on *direct sampling* in which the number in the second sample is predetermined. Alternatively, *inverse sampling*, where the number of marked birds to be recaptured is predetermined, has the advantage of giving an unbiased population estimate and variance in cases where the number of recaptures is small:

$$P = \frac{n(a + 1)}{r} - 1$$

An approximate estimate of the variance of this is given by:

$$\text{var } \hat{P} = \frac{(a - r + 1)(a + 1)n(n - r)}{r^2(r + 1)}$$

Further details of this group of methods are given in Southwood (1978).

Example of estimating population size in the Pheasant using the simple Lincoln index

Consider a simple example in which, during the winter of 1984/85, 27 male and 70 female Pheasants were marked using plastic back-mounted fin-tabs, each bearing an identifying number. Trapping was conducted continuously each day and as more birds were marked, the number of recaptures increased. The traps were not moved and were baited with grain to attract birds into them. Although the primary aim was not to estimate

Box 6.3 cont.

population size from these data, but to investigate behavioural trends in spring, it is useful to consider the application of the Lincoln index method in estimating population size. A comparison will then be made with that for the spring population using the marked birds observed from a vehicle in daily morning and evening watches.

Catches made after the 70 females were marked included 62 marked birds out of 66 birds caught.

$$P = \frac{70 \times 66}{62}$$

= 74 females = size of winter population.

Catches made after the 27 males were marked included 21 marked birds out of 24 caught.

$$P = \frac{27 \times 24}{21}$$

= 31 males = size of winter population.

This gives a total winter population of 105 birds.

The main limitations of this method are assumptions 1–6 in Box 6.2. Those of population closure, no mortality and equal capture probability, are most important. Pheasants are relatively sedentary during the winter although immigration and emigration occur during late winter and early spring (and in some cases throughout the winter). Further, the period over which capture took place was too long and some mortality due to Fox predation was known to have occurred. Some individuals, particularly hens from the same harems, were trap-happy and usually a greater proportion of hens from the same harem were caught than would be expected by chance, since they feed together during the winter and early spring period. For these reasons the use of the Lincoln index is not recommended in this particular case.

The spring censuses of Pheasants on the study area were used to estimate a spring population size for, approximately, a 1-month period by mapping all birds seen on successive morning and evening visits throughout mid April–mid May. The position of marked and unmarked males and females (together with identifying code for those marked) were entered on a map. The proportion of males and females on the census that were observed tabbed (CT) for the whole of the spring observation period are shown below. The total number (TT) of males and females tabbed on the study area during the catching period is also shown. Simple Lincoln index calculations permit the estimation of population size of the sexes separately. In this case recapture events are actually resighting events.

The Lincoln index method used on resighting data for Pheasants rather than on recapture data:

Census birds tabbed (CT as percentages)	50 males, 59 females
Total birds tabbed (TT)	27 males, 70 females
Estimated population $\left(\dfrac{100}{CT} \times TT\right)$	54 males, 119 females

From this it can be suggested (with the proviso of population closure, no mortality and equal captive probability) that both the female and male 'populations' have been

Box
6.3
cont.

added to from late winter to early spring. This is one of the many benefits of having birds marked with visible tags which allows 'sampling' to be observational without capture.

To summarise the Pheasant example, there are various assumptions with this method, most important of which are (1) that the marked birds must freely mix within the extant population; (2) all birds should have an equal probability of being seen; (3) tags must be permanent and must not fall off; and (4) no mortality should occur from the period of capture and marking to the period of resighting. These correspond to assumptions 4, 2, 5 and 1, respectively, in Box 6.2.

The du Feu method of analysing capture–recapture data.

Box
6.4

Du Feu *et al.* (1983) present a method for making 'spot estimates' of populations in which the number of new birds (*N*) and the number of recapture events (*R*), over *S* captures, are used. The number of new birds (*N*) is the number of birds captured for the first time in the session. If a bird is captured four times in a session it contributes 1 to *N* and 3 to *R*. The equation is:

$$1 - \frac{N}{P} = \left(1 - \frac{1}{P}\right)^S$$

where *S* is also equal to *N* + *R*. The equation has no explicit solution so *P* has to be found iteratively.

One begins by making a sensible guess. Then, with successive sessions, and by inputting *N* and *R* from them, one can use a program to make a running estimate of the population. A graph of running estimate (*P*) against capture event (*N* + *R*) will show if the estimate is stable, which can indicate when the trapping should be stopped, since more captures would not alter the estimate. Such a plot would also indicate whether the population is closed in that, for example, immigration would give rise to a decrease in recapture events and an increase in new birds. An example of how the estimates are refined as the number of recaptures increases is shown in Box 6.5.

The standard error of *P* is:

$$SE_P = \sqrt{\frac{P}{e^{\left(\frac{N+R}{P}\right)} - 1 - \frac{N+R}{P}}}$$

Example of the use of the du Feu method on the Pheasant data

The values of *N* (the number of new birds caught) and *R* (the recapture events) for the same session i.e. winter period, gave du Feu population estimates of 51 ± 12 for males and 89 ± 6.6 for females, with a total population estimate of 132 ± 9.7 birds. Running estimates using the du Feu method could also be made for successive capture events through the catching period. The Lincoln index method gave a total population estimate of 105 birds, compared to 132 birds using the du Feu method.

Box 6.5

The du Feu method for calculating population size.

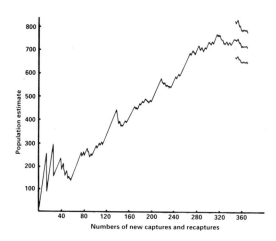

In this method, each bird captured, whether it is a new bird or a recapture, enables a new population estimate to be made. If each successive estimate is plotted against the capture event it is possible to follow fluctuations in the estimate, and to identify the point at which it settles down. Confidence limits are also calculated and plotted (from du Feu *et al.* 1983).

Box 6.6

Examples of du Feu estimates of population size against number of capture events.

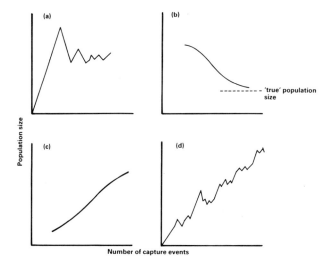

(a) Population closed; (b) population with emigration; (c) population with steady immigration; (d) population with step-wise immigration (from du Feu *et al.* 1983).

Population indices derived from the UK Constant Effort Sites Scheme.

Box 6.7

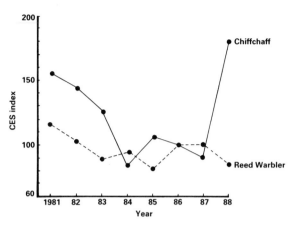

The base year of the index is 1986, and Chiffchaff and Reed Warbler indices are shown from 1981 to 1988 (from Peach and Baillie 1989). Reed Warbler populations have remained stable and Chiffchaff populations have increased, for unknown reasons.

Productivity monitoring using the BTO Constant Effort Sites Scheme.

Box 6.8

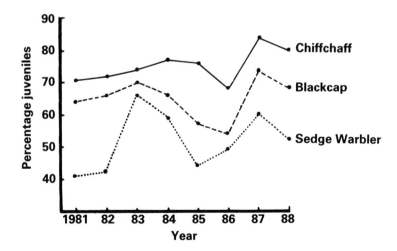

Changes in the percentage of juveniles caught under constant trapping effort at specific sites are shown for Chiffchaff, Blackcap and Sedge Warbler for 1981–1988 (from Peach and Baillie 1989). These show broadly similar trends.

Box 6.9

Importance of standardisation in netting operations.

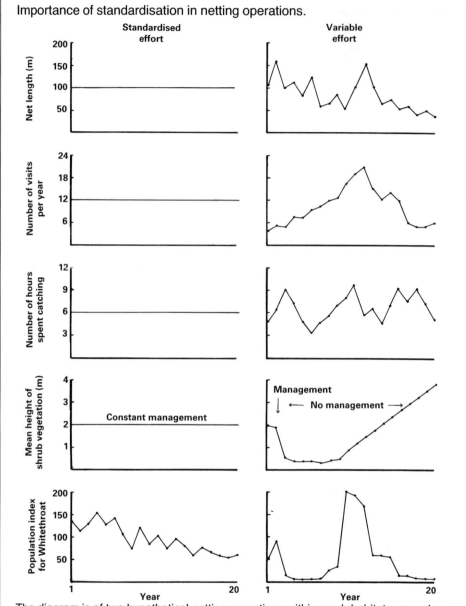

The diagram is of two hypothetical netting operations within scrub habitat—one where capture effort has been standardised and the other where capture effort has been variable over the past 20 years. Population indices for Whitethroat calculated from captures are different, yet the indices for Whitethroat in the standardised-effort system were used to calculate those in the variable-effort system. The standardised-effort system shows a decline in numbers, the variable-effort system does not. We should put more trust in the index trend for the standardised-effort system whereas nothing useful could be said about population trends from the results from the variable-effort system.

Capture per unit effort where ringing effort is not known—correcting for annual variations in ringing effort.

Box 6.10

To correct for annual variations in ringing effort, the numbers of individual species ringed each year can be transformed into numbers of each species ringed per 100 000 birds ringed, referred to as the standardised annual totals according to the formula:

$$Y'_n = \frac{Y_{n-2} + 2Y_{n-1} + 4Y_n + 2Y_{n+1} + Y_{n+2}}{10}$$

where Y_n is the annual total of birds ringed in year n, and Y_h is the weighted running mean of year n.

Variation in ringing effort from year to year is difficult to estimate since few ringers or observers record the total length of net they use, the total number of hours they spend ringing, or the variation in weather during ringing sessions. To correct for this variation the total number of birds of all species per year is used, and individual species totals are adjusted accordingly to remove any increasing or decreasing trend in totals ringed which, it is assumed, is due not to increases in bird populations but to an increase in ringing effort. An example of the use of this method to identify population trends in birds ringed in Sweden is given in Österlöf and Stolt (1982).

Comparison of population indices produced by different methods.

Box 6.11

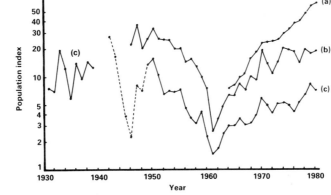

This figure presents the results of three methods of population indexing for the Stock Dove in Britain since 1930.

(a) From Common Birds Census; (b) from ringing totals; (c) from nest record card totals (from O'Connor and Mead 1984). Dashed lines are periods with limited nest card returns.

Box 6.12

Example of the use of the simple matrix model.

Let us assume we have the following data for a hypothetical population in which we know the current size of the population because we have censused each age cohort:

Age (years)	Fecundity (young produced)	Survival rate	Population size
0	0	0.2	650
1	6	0.5	320
2	10	0.8	65
3	5	0	30

Hence, no offspring are born to adults of between 0 and 1 year, six are born to those of 1–2 years, etc. Survival is highest in the age category 2. These data are then used to construct (1) a transition matrix (T) and (2) a column vector of the numbers in each age category (A) in the following way:

$$\begin{bmatrix} T & & & \\ 0 & 6 & 10 & 5 \\ 0.2 & 0 & 0 & 0 \\ 0 & 0.5 & 0 & 0 \\ 0 & 0 & 0.8 & 0 \end{bmatrix} \times \begin{bmatrix} A \\ 650 \\ 320 \\ 65 \\ 30 \end{bmatrix}$$

When the transition matrix (T) and column vector (A) are multiplied, we get:

$$= \begin{matrix} (0)(650) & + (6)(320) & + (10)(65) & + (5)(30) \\ (0.2)(650) & + (0)(320) & + (0)(65) & + (0)(30) \\ (0)(650) & + (0.5)(320) & + (0)(65) & + (0)(30) \\ (0)(650) & + (0)(320) & + (0.8)(65) & + (0)(30) \end{matrix}$$

$$= \begin{matrix} 2720 \\ 130 \\ 160 \\ 52 \end{matrix}$$

This new column vector is the number of individuals in the population of the next generation according to the four age categories. Matrix models allow us to condense the complexities of age-specific schedules into a simpler form. One disadvantage of the model outlined here is that no account is made for density-dependent survival or fecundity, which have been shown to exist in many bird species. More recent developments in matrix modelling have addressed this problem. The reader is referred to Begon and Mortimer (1986) for further details of matrix models.

Major considerations for undertaking a radio-tracking study.	**Box 6.13**

Consideration	Comment
Activity of the bird	Will breeding or feeding behaviour damage the transmitter?
Weight	How big is the bird relative to its probable range in the habitat?
Weight of transmitter in relation to body size	Should be less than a notional 5% of the bird's body weight
Harness design	Will this affect its behaviour? Tail mounted, back mounted, collar mounted, leg mounted?
Range	What is the habitat; will this reduce or improve expected range of transmitter?
How many birds to mark	Logistics of observer moving between radio-location points. How many birds can be tracked comfortably at any one time?
Special transmitters	To monitor bird activity e.g. flying or resting, upending by ducks, depth measurements in diving birds, physiological measurements e.g. heart rate
How much sampling	How many radio-locations and over what period? e.g. 30 locations over 10 days for Pheasants will define their home range adequately for analysis during this period
How to obtain data in the field	Triangulation from a vehicle, on foot; use of a data logger linked to a computer database; automatic triangulation from a fixed location
Analysis of home range from radio-location	Minimum polygon area; probabilistic methods; harmonic mean; Kenward (1987) 'Ranges' suite of programs

Box 6.14

Comparison between radio-tracked and visually marked Pheasants.

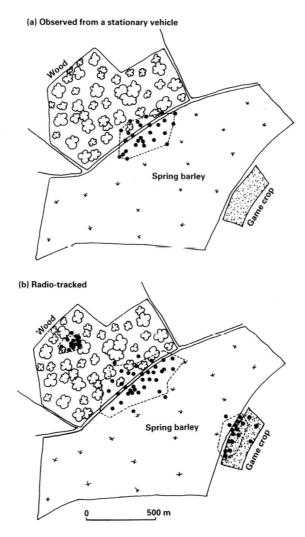

(a) Observed from a stationary vehicle

(b) Radio-tracked

0   500 m

Hypothetical example of the value of radio-tracking data when compared with that obtained from the same bird which was back-tabbed. In (a) the data obtained from a census from a stationary vehicle are restricted. If the Pheasant (●) enters the wood, or moves a long distance, the census might fail to pick it up. In (b) the same bird is radio-tracked, and this identifies an area around the nest-site within woodland which was previously unrecorded, and a feeding area in an old game crop a long distance from the main part of its range. Consequently, if the stationary vehicle census data were used to say something about habitat selection by the bird, a biased picture would be presented.

Assumptions

1. The population is either closed or immigration and emigration can be measured or calculated

A closed population is one where there is no immigration or emigration of birds during the period of population estimation. There should also be no births or deaths within the sampling period unless allowances can be made for them.

Likely causes of violation: immigration, emigration, births or deaths occurring at the time of the study may be exacerbated by captures being taken over a long period.

Reducing the problem: reduce time intervals over which captures are made and conduct captures at a time of the year when migration and recruitment of young birds to the population are not occurring. Sampling should be at discrete time intervals and the time involved in taking the samples should be small in relation to the total time.

2. There is equal probability of capture in the first capture event

All individuals of the different age groups of both sexes should be sampled in proportion to that in which they occur in the habitat. Therefore, all individuals of the different age groups should be equally available for capture irrespective of their position in the habitat.

Likely causes of violation: there may be part of the population that is never captured because the individuals concerned are trap-shy or they cannot be sampled in certain habitats. Alternatively some birds may be trap-happy.

Reducing the problem: if it is possible that the probability of capture differs between the sexes or other 'sub-groups', population estimates should be derived independently. For example, female pheasants are both trap-happy and more likely to be captured than males whilst living in harems prior to egg-laying. The regular movement of traps within the study area reduces the over-sampling of these individuals. It may be desirable to use traps to catch the first sample and to apply highly visible markings so that the second sample can be obtained by re-sighting rather than capture. This has been used successfully on Mallard (United States), Pheasants (United Kingdom) and Willow Ptarmigan (Red Grouse) (Scandinavia).

3. The marked birds should not be affected by being marked

Likely causes of violation: marks significantly affect behaviour e.g. neck collars have been reported to contribute to starvation in Snow Geese. Higher re-sighting probabilities have been suspected for patagial-tagged, back-tagged and dyed birds, than for unmarked birds, in studies in the U.S.A. Increased predation of birds tagged using patagial markers has also been recorded.

Reducing the problem: be aware of these effects and remedy by changing the marking system.

4. The population should be sampled randomly in subsequent capture events

Likely cause of violation: individuals do not mix randomly and perhaps change the area they use, biasing future capture.

Reducing the problem: this is a difficult problem to overcome. More traps and more field observations are desirable.

5. The marks should be permanent

Likely causes of violation: marks fall off or become unreadable at a distance.

Reducing the problem: be aware of unsuitable designs which have a high failure rate. Double mark where necessary (e.g. a leg ring or band as well as a back-tag). On future capture worn or lost markers can be replaced.

6. All marked individuals occurring in the second or subsequent samples are reported

Likely cause of violation: this assumption is generally relevant to experiments in which the second sample is based on ring or band recoveries made by the general public (e.g. hunting recoveries).

Reducing the problem: the total number of recoveries for use, for example, in Lincoln-index model estimates requires an estimate of the reporting rate, i.e. the proportion of recovered rings or bands that is reported. Reporting rate has been estimated using either additional information on the number of recovered rings obtained from hunter questionnaire surveys, or reward studies in which some rings or bands are marked with a message that a reward is offered for their return.

7. Capture probabilities are assumed constant for all periods

Likely cause of violation: capture probabilities will vary from one capture event or sampling period to another, as a result of weather factors and possible changes in sampling effort.

Reducing the problem: conduct sampling under the same weather conditions wherever possible, and use the same sampling effort, e.g. length of capture net, number of traps, etc.

8. Every bird in the population has the same probability of being caught in sample i

This assumes that the bird is alive and in the population during sampling period i.

Likely causes of violation: age-specific and sex-specific variation in capture or habitat use.

Reducing the problem: sample the different age, sex and other 'sub-groups' independently.

9. Every marked bird in the population has the same probability of surviving from sampling periods i to $i + 1$

This again assumes that the bird is alive and in the population immediately after the time of release in sample i.

Likely causes of violation: age-specific and sex-specific variation in survival, e.g. higher mortality of female waterfowl (as shown in studies in the USA for example), caused by differential predation.

Reducing the problem: sample the different age, sex and other 'sub-groups' independently.

10. Every bird caught in sample i has the same probability of being returned to the population

Likely causes of violation: differential age, sex and individual stresses causing mortality during handling.

Reducing the problem: be aware of experiences and keep handling time to a minimum.

11. All samples are taken instantaneously such that sampling time is negligible

Likely causes of violation: this assumption is never strictly met.

Reducing the problem: as previously, keep the sampling period short.

12. Losses to the population from emigration and death are permanent

Likely causes of violation: birds that have emigrated return to their natal site, showing high site fidelity (common in birds).

Reducing the problem: use capture and re-sighting efforts. Radio-telemetry is a useful technique in distinguishing between dispersal movements and emigration, and direct mortality.

13. Population closed to recruitment only

Likely causes of violation: recruitment occurs but goes unnoticed.

Reduction of problem: choose the time of year when recruitment does not occur. This has been successfully achieved in American Woodcock, using a model known as the 'death but no immigration model'.

Estimating population size

There are many methods for estimating population size, and only a few are given here. They are (1) the simple Lincoln index, (2) the du Feu method, and (3) other methods in which there are more than two sampling occasions.

1. The simple Lincoln index

The simplest Lincoln index model is based on one capture and one recapture only, i.e. a two-sample case, although computations for a multi-sample case in which a number of sub-samples are taken are described below. Variants of the method have been derived to allow for losses (emigration and death) or gains (immigration and birth) to the population. Provided the conditions listed under model a in Box 6.2 are satisfied, the total population, size P, can be calculated from the simple Lincoln index as shown in Box 6.3. The method is most appropriate for studies of colonial species which cannot be counted directly such as a seabird colony (e.g. Storm Petrel) or warblers in a reedbed. In both cases the boundary of the habitat is known and can be defined. It is perhaps least useful for populations that are continually changing in number, or for birds in more complex habitats such as gardens.

2. The du Feu method

The standard methods of mark–recapture analysis, such as Jolly's (1965), require repeated samples being taken at fixed intervals from which measures of immigration and emigration may be estimated. The du Feu method (du Feu *et al.* 1983) is particularly useful for estimating the population of a species, such as a warbler in a wood, being ringed during a single session e.g. day, week or season, depending on the mobility of the species concerned. If the species is sedentary, the session can be a winter or other field-season measure. Details of the method are given in Boxes 6.4 and 6.5.

Plots of typical running du Feu estimates are shown in Box 6.6 for four scenarios: (a) for a closed population wide fluctuations are rapidly damped to a relatively stable value; (b) for declining populations the estimate slowly declines, but is always greater than the true population—the estimate can never be less than the number caught (N), although in a rapidly declining population there may be fewer birds present at the end of the season than have been ringed; (c) for smoothly increasing populations the estimate increases smoothly; (d) for a stepwise increasing population the estimates consist of a series of steps.

The conditions and limitations of the method refer to assumptions under class b in Box 6.2. In particular the method assumes population closure, equal probability of capture, no modification to behaviour caused by trapping, and additionally (assumption 11 of Box 6.2), negligible handling time of the birds relative to the study period. In respect of the assumption of population closure, trapping should be carried out for a long enough period to obtain a reasonable number of recaptures, but not so long as to risk immigration and emigration. In isolated small habitats it is possible that the whole population is being sampled, whereas in large, uniform habitats the estimated population may be an accurate measure of the catchment area population, but that area may not be known. Where individuals are being caught at an exploited resource, e.g. a water source, it is the usage that is being estimated, not the population.

Underhill and Fraser (1989) have developed a Bayesian analogue of the du Feu estimate and have used the method to estimate the number of Malachite Sunbirds at a flower food source in South Africa. The assumptions are identical to those of the du Feu method but the Underhill and Fraser method is computationally simpler.

Bayesian methods assume some prior knowledge about the maximum size of the population being trapped whereas the du Feu method does not require prior knowledge or assumptions of maximum population size. As new birds are caught the method calculates the new probabilities of there being x birds in the population. Each unringed bird captured shifts the probability distribution to the right, increasing the estimated population size; each retrapped bird shifts the probability distribution to the left, leading to a decline in the estimated population size.

The method has a number of advantages over that of du Feu, and is likely to be used more extensively in the future, in preference to the du Feu method. First, it provides more realistic confidence limits because they are based on an exact probability distribution. Second, the method is better than that of du Feu for small populations (<100 birds). The main disadvantage of the method is that if the initial guess of the maximum population is too low, the population estimate will approach its asymptote at the far right of the plot, therefore it is important not to make a serious underestimate from the outset. There is no harm done to the estimate of population size if the maximum population is overestimated however.

3. Other methods with more than two sampling occasions

For useful reviews see Nichols *et al.* (1981) and Pollock (1981). These methods are generally multi-sample Lincoln index type models and the sampling regime is similar to a Lincoln index two-sample case in that birds are captured during an initial sampling period, marked and returned to the population. A second sample is then taken (e.g. on the following day) and recaptures of marked birds are noted. New captures are also given marks and all birds are returned to the population. This procedure is repeated for K sampling periods. The main difference between these methods and the Lincoln index two-sample method is that each bird must be given an individual mark e.g. serially numbered leg ring. The models used to describe recapture data from K sample studies require complete capture histories. The probability distribution for the set of possible capture histories is then expressed using a multinomial model treating population size and capture probabilities as parameters. Examples of models for closed populations are described in Otis *et al.* (1978). Assumptions of these types of models relate to class b in Box 6.2. For short-term studies, closed population models allow for unequal catchability of individuals (see Pollock 1981).

Models that take account of population gains and/or losses, i.e. for completely open populations have been developed by Jolly (1965) and Seber (1965). These are denoted the Jolly–Seber stochastic models for completely open populations. Both population size and survival rates are calculated. The model has not been extensively used because of violation of a number of the assumptions presented under class c in Box 6.2. Further, a better survival model has been developed by Clobert *et al.* (1987), although population size estimation is not a feature of their model. For long-term studies, open population models that assume equal catchability are used (see Pollock 1981). Such models allow estimation of survival and birth rates as well as population sizes.

Mark–recapture models for populations that are open to both gains and losses, as described in the last paragraph, have the greatest potential applicability to studies on bird population dynamics. Models do, however, exist for populations that experience only losses, and no gains, and which are subject to the assumptions in class d of Box 6.2. See Nichols *et al.* (1981) for further descriptions.

Methods based on catch per unit effort

Methods based on capture per unit effort rely on standardising effort of capture or observation. The aim is to estimate a species abundance by making the effort by which the abundance data are obtained constant. The main assumption is that the standardised design yields unbiased data from which to calibrate population size or an index thereof. Seber (1973) provides a review of the method for closed and open populations.

1. Where ringing effort is standardised

By controlling i.e. standardising the effort invested in catching and marking, detailed studies of (1) population size, (2) productivity and (3) survival can be made. The Constant Effort Sites Scheme (CES) of the BTO aims to collect data to investigate changes in these variables by using a series of constant net-sites worked on each of 12 standard visits spread between May and August. The following are investigated.

(1) Index of population change—changes between years in the numbers of adults captured.

(2) Productivity—the ratio of juveniles to adults captured late in the breeding season.

(3) Survival—between-year retraps of ringed birds.

The same ringing and netting sites, with the same length and type of net, are used each year. No other netting is carried out within 400 m of the net-sites. Where possible, sites within a single major habitat are preferred. Continuation of the study is important in order to minimise fluctuations in population indices resulting from changes in the habitat or geographical composition of the sample, and to allow comprehensive survival estimates to be calculated.

Netting is conducted for a set time of about 6 hours per visit and no tape lures or baits are used (as in other studies) to attract birds to the site. Habitat is recorded on 1:2500 scale maps. The validity of measuring population changes by mist-netting is assessed by comparing results with those from point counts (see Chapter 5), which are conducted at the site in the time between visiting individual nets.

The results are interpreted by (1) calculating changes in the number of adults caught between years using the same sites, to provide a measure of change of the size of the adult population; (2) calculating the proportion of young birds caught on the same sites as an estimate of productivity in the post-fledging period; (3) calculating survival rates using the SURGE routines described in Clobert *et al.* (1987).

Box 6.7 shows an example of a CES population index in which the base year of 1986 is taken as 100. Box 6.8 shows an example of productivity monitoring from CES sites in which the proportion of those caught represented by juveniles is plotted each year. Significant changes in the population index and proportion of juveniles can be calculated.

Another example of a programme in which ringing effort and habitat is standardised is that of the 'Mettnau-Reit-Illmitz' (MRI) scheme (Berthold *et al.* 1986). The three sites are in central Europe (S and N Germany and E Austria) and trapping of passage migrants has taken place from the end of June to the beginning of November since 1974. Trapping takes place daily, for the same length of time per day, using the same netting system and the same length of nets at the same number of net-sites. Vegetation is trimmed to maintain consistent profiles between years so as to halt the effects of successional changes in habitat. Five regression models were used to analyse correlation coefficients for regressions of total numbers of individual species captured against year, 64% of the coefficients being negative implying population declines.

Box 6.9 shows a diagram of two hypothetical netting operations within scrub habitat. At one site the same net length is used, visits are made every week, for the same length of time every day, in similar weather, and vegetation profiles are trimmed to keep the habitat at the same successional stage. The other site uses variable net length, different observers, variable work day length and variable number of visits through the season, and the habitat is succeeding to woodland. Population indices for Whitethroat calculated from captures from the two sites are not very similar. Which one should we trust? Obviously, the data collected under standard effort is more trustworthy.

Ormerod *et al.* (1988) used a constant length of mist-net and constant catching effort to sample Dipper, Kingfisher and Grey Wagtail along river systems in Wales. The effective width of the river sampled was also recorded. The results for Dippers were calibrated with the known abundance. Mean catch per hour was correlated with the number of birds per 10-km stretch

($r = 0.99$), as was the mean catch per hour standardised per 10-m net ($r = 0.96$). The study concluded that standardised ringing along rivers can be used in population monitoring and suggested that the method may be effective in assessing annual change in breeding success, post-fledging survival and overwinter survival.

2. Where ringing effort is not known

Annual ringing totals can be used to analyse population trends. The main assumption is that there is a relationship between changes in annual totals ringed and the abundance of a species when corrections are made for ringing effort. Box 6.10 shows how to correct annual variations in ringing effort for individual species, and therefore how to relate this to variations in a species' abundance.

A similar approach has been used for data for the Stock Dove in Britain (O'Connor and Mead 1984). The number of nestling Stock Doves ringed annually per 1000 nestlings of all species ringed nationally during 1931–80 was calculated. The all-species totals alter with ringing effort, but, with a large number of species involved they are likely to average out species-specific fluctuations. In this analysis information was further compared to annual ratios of nest record cards submitted to the BTO (dating back to 1930), as well as the Common Birds Census Index (dating back to 1962). Box 6.11 shows the agreement between the three methods of population indexing since the early 1960s, and can usefully suggest population trends prior to censusing (i.e. pre-1962).

Further developments of studies involving capture

There are two more specialised procedures which involve the use of catching and marking techniques but which fall outside capture–recapture and capture per unit effort categories. These are (1) the use of matrix models in which productivity, population sizes and survival estimates are used to predict future numbers of a species (i.e. using some of the techniques previously described), and (2) the use of radio-telemetry in the special case of studying the change in location of an individual bird and its use in calibrating with more extensive but more cheaply obtained data.

1. Modelling with birth and survival rates—the matrix model

Survival and fecundity (leading to productivity) are often age-specific in certain species. The descriptions given below present one method of modelling data obtained from ringing recoveries when fecundity of age cohorts is known. The matrix model was developed by P.H. Leslie in 1945.

There are three requirements for using the matrix model: (1) the number of individuals of each age category in the population being studied (P); (2) the age-specific fecundity (B); and (3) the age-specific survival rates (S), derived, for example, with the use of the SURGE program (Clobert *et al.* 1987). An example of the matrix model is given in Box 6.12.

2. Radio-telemetry

This section is meant only to give an overview of the major considerations to be addressed when embarking on a radio-tracking study. For a fuller description of the technique and analytical routines see Kenward (1987). The radio-tracking system usually consists of the transmitter, which is fixed to the bird, a multi-channel receiver which detects the signal emitted by the transmitter, and one or two hand-held portable directional or fixed antennae. Box 6.13 lists the main considerations and provides additional comment. Box 6.14 shows the value of radio-tracking in being able to pick up movements of the bird to new areas: observations of the tabbed Pheasant from a vehicle were more restricted and did not pick up the bird within woodland or when it moved some distance to feed in a game crop. The use of radio-telemetry in studying bird distribution at the habitat-scale, and habitat selection, is described in Chapter 9.

Summary and points to consider

Is it necessary to catch and mark the individuals in order to satisfy the objectives of the study? How is this related to bird counting?

Design of marking method, catching protocol and time constraints must be addressed.

If population size is to be estimated using catching and marking methods, are any of the assumptions of the analytical method violated? Can any violations be dealt with?

Consider a number of methods for achieving the same product, as in the Stock Dove example where long runs of data are available. Are these products similar, i.e. do the methods provide a consistent interpretation of, for example, changes in population size?

Wherever possible consider standardised and replicated procedures for collecting data from marked individuals. Reference, for example, to the Constant Effort Sites scheme is important here.

7

Counting Individual Species

Introduction

For some species the 'standard' methods of counting breeding populations such as territory mapping (Chapter 3), point counts (Chapter 5) or line transects (Chapter 4) are not particularly successful. Reasons for this may be a low breeding density (e.g. raptors), a secretive nesting habit (e.g. ducks, many waders), crepuscular or nocturnal ecology (e.g. owls, nightjars), or semi-colonial and colonial nesting behaviour (e.g. herons, Rooks). For such species other methods have been developed which aim to produce an index of the breeding or non-breeding population and thereby facilitate comparisons of populations between years and sites. In some cases these methods are modifications of territory mapping or transect methods, but two other types of counting are commonly used: direct and indirect counting and 'look-see' counting. These are explained below.

Direct and indirect counting

For direct counting a suitable vantage point is selected and all visible birds are counted. The method is very useful when all birds can be easily seen, e.g. at raptor migration bottlenecks, places where waders breed on small islands, wader roosts, or smaller seabird colonies (also see Chapter 8).

With direct counting, a high level of accuracy at a given time may be possible, but there are several ways in which the results can become biased. The most important are a failure to ensure even effort and coverage between sites or years, resulting in data that are not comparable. Other factors such as the weather during the counting, the people undertaking the counts and whether the naked eye, binoculars or telescopes were used will all influence the accuracy and comparability of the counts.

Indirect counting relies on counting signs of bird activity (droppings, burrows, etc.). This may be the only suitable method for particularly secretive (e.g. rain-forest pheasants) or non-visible birds (e.g. those nesting in burrows: Chapter 8). Theoretically, indirect signs of birds might be calibrated to produce indices of population level. This may be possible when counting nesting burrows. However, when counting droppings, both the number of birds and the length of time they have spent in that particular area will influence the number of droppings recorded. Hence, the number of droppings counted per unit area gives an index of bird usage, not necessarily population level (see Owen 1971).

Look-see counting

The look-see method relies on a prior knowledge of the habitat-preferences of the bird, and, more than any of the previous methods, requires the observer to 'know' well the bird concerned. Potentially suitable habitat for the species of interest is identified (1) from a map of the area, (2) from an aerial photograph of the area, or (3) through contact with local experts. Once suitable areas are identified, a programme of site visits is arranged at the appropriate time and using appropriate methodology to count any birds present. From such counts, population estimates are made (Box 7.1).

Such counts will become biased if the area covered or the effort put into the counting varies over time. There is also a general tendency for more birds to be discovered over a period of years as the study area becomes better known; hence counts may tend to increase over time.

Data on the habitat preferences of many birds, which can be used as the basis of Look-see counts, are available in Cramp & Simmons (1977 onwards: Western Palaearctic), Brown *et al.* (1982 onwards: Africa) and Palmer (1962 onwards: North America).

Methods for counting different groups of birds

This section presents methods for counting populations (mainly during the breeding season) of a selection of bird-groups where standard methods do not work well. Examples are arranged in taxonomic order and most are taken from the British Isles. However, it should be possible to adapt methods to count similar species throughout the world.

In all cases the counting methods are quite specific and hence it is difficult to compare results between species. Such counts are useful in studies that focus on one or a small number of species and are not appropriate for community studies where there may be several counting methods in use at the same time.

1. Divers

Look-see methods are employed to count divers breeding at low density in remote areas.

For *Red-throated*, *Black-throated*, and *Great Northern Diver* the counting unit is the adult bird. The best counting period is late incubation or early fledging as all territories and nests are by then well established (mid May to late June in Britain). Waterbodies within the study area are identified from maps and then visited twice during the survey period, with the second visit not less than 2 weeks after the first. Two visits assess site-occupancy more fully than one as birds may be feeding away from the site on one of the visits.

On each visit the whole loch should be systematically scanned from suitable vantage points to detect adult divers on the water (Box 7.1), and all known nest-sites checked. Excessive walking along the shoreline, or searches for nests, should be avoided as they are extremely time-consuming and may disturb the nesting birds.

Detailed studies in northern Scotland (Bundy 1978; Campbell and Talbot 1987) used the following criteria to indicate a breeding territory: (1) chicks seen, (2) nests with eggs seen, (3) pair of adults on loch on two visits, (4) pair of adults on one visit but single adult on the other visit, (5) single adult on loch on both visits.

2. Grebes

Grebe species are often widely dispersed over suitable wetland sites. Methods for counting an obvious and secretive species are presented below.

OBVIOUS SPECIES

For the *Great-crested Grebe* the counting unit is the adult bird. A recent study in Denmark (Woolhead 1987) indicates that an accurate population estimate can be obtained by counting the number of adult Great-crested Grebes on a waterbody several times up until the start of the first nest-building attempt (April in Britain: Hughes *et al.* 1979). Such counts should not be affected by either changing visibility as vegetation grows or failed breeders moving between lakes.

Woolhead (1987) showed that a reliable estimate of the number of breeding pairs could be produced by halving the average number of adults obtained from the counts. However, overestimation is possible on large lakes as non-breeding birds tend to congregate and might be regarded as breeding pairs.

SECRETIVE SPECIES

The *Little Grebe* is highly secretive, and in addition pairs often make several breeding attempts, moving between lakes when doing so; thus breeding populations are difficult to assess. Vinicombe (1982) recommends making several counts of a study site over the first breeding attempt of the year (April to May in Britain). On each visit all sight and calling records are marked on a site-map. Often the only clues to the presence of breeding pairs are the 'trilling' calls.

An estimate of the number of breeding territories can be produced using methods outlined in Chapter 3. Alternatively, a single visit can be made during the first breeding attempt and all birds recorded. This will produce a population index which will allow comparison of breeding population levels between years.

Box
7.1

Look-see surveys of species in remote areas.

(a)

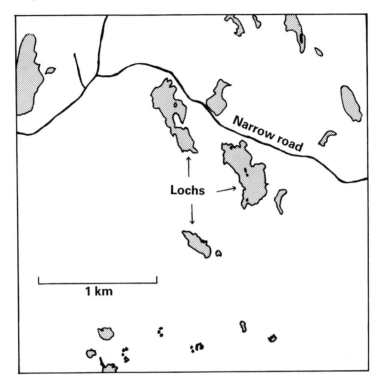

(b)

(a) When counting divers on lochans in Scotland all the lochans in the study area must first be located and a programme of site visits planned to each of them, giving equal coverage effort.

(b) At each site the whole surface of the water must be scanned slowly and carefully from side to side. Too little time spent at each site will lead to birds being missed if they are behind vegetation, during periods of poor visibility, etc. To obtain accurate results even the most remote and inaccessible lochans must be counted using the same amount of effort as the most accessible sites. Otherwise results might reflect the ease of reaching sites rather than the population of birds in the area.

Box 7.2

Census form for an obvious breeding waterfowl species.

BTO/WWT/SOC MUTE SWAN CENSUS 1990: BREEDING PAIRS

County (England & Wales) or District & Region (Scotland)_____ AVON

Please mark on the grid the positions of all pairs and nests found, using the following symbols to represent the state you recorded on your last visit:

Territorial pair .... ✕
Pair with nest .... ○
Pair with brood .... ●

Against each symbol write the letter used for the site code on the other side of this form.

Please shade any parts of the 10 km square that you were unable to cover.*

What is your best estimate of the number of pairs in the shaded part? What is your reason for this belief?*

O

No SUITABLE HABITAT

Please write your name, address, and phone no. here:

J. SMITH   654321
1 FIELD ROAD
BRISTOL
AVON.   BS1 1AA

*Note: estimates are much less useful than proper coverage.

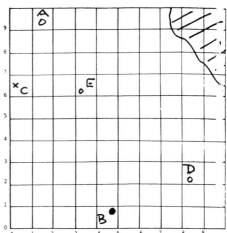

As soon as possible after 31 May, please return to your local organizer.

Your local organizer is:

SIMON DELANY
WWT, SLIMBRIDGE, GLOS.
GL2 7BT   (0453-890333)

MUTE SWAN CENSUS 1990: BREEDING PAIRS: OBSERVATIONS

Site code	Location	Grid ref.	Habitat	Dates and observations (Dates as 00/0 please)
A	LITTLETON BRICKPITS, NR. WICK	716092	GRAV. PIT	03/4 T, 15/4 N, 01/5 N
B	ALVESTON RES.	747007	RESERVOIR	03/4 N, 01/5 N, 27/5 T
C	HILLSIDE LAKE THORNHAM	701065	LAKE	03/4 T
D	LITTLE AVON NR. THORNBURY	784021	STREAM	15/4 N, 07/5 D
E	R. SEVERN, FROME	733062	RIVER	01/5 N

Shooting bag records as indices of population level. **Box 7.3**

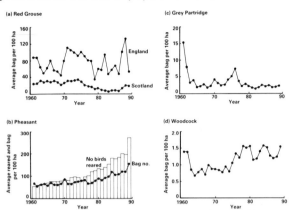

Historical records of shooting bags of Red Grouse, Pheasant, Grey Partridge and Woodcock are available from a wide sample of sites throughout Britain (from Tapper 1989). These provide valuable information on population trends over extended periods. However, other factors need to be considered for each species before the true meaning of the graphs can be assessed.

(a) Average bag records of Red Grouse per 100 ha of moorland from several sites in England and Scotland. Numbers shot in England show cyclicity, with 1989 being a poor year. However, there are several other facts needed before the graph can be fully evaluated; these are the number of persons shooting, changes in the shooting equipment, periods of adverse weather, changes in the law on shooting or gamekeeping, etc. Such factors will influence the numbers of birds shot and hence the shooting bag records obtained.

(b) Average bag records of Pheasant per 100 ha. Taken alone this graph might tend to suggest an increasing natural population. However, the bar charts on the graph present the number of young birds artificially reared and released into the wild. Considering these data, it now appears that the number of birds shot has, in fact, not increased in line with the number released. An increasing number of birds being released also implies increasing interest in Pheasant shooting and thus increased shooting pressure. Much additional background data on trends in Pheasant rearing, Pheasant shooting and the status of the wild population need to be available before this graph can be fully interpreted.

(c) Average bag records of Grey Partridge per 100 ha. The graph shows a dramatic decline in numbers shot in the early 1960s, followed by an extended period when the number of birds shot has remained low and relatively stable. The population decline in the 1960s is believed to be due to the increased usage of pesticides from that time, reducing the food supply for Grey Partridge chicks (Potts 1986). However, to interpret the graph fully, the level of shooting pressure over time and the number of birds released over time would need to be known.

(d) Average bag records of Woodcock per 100 ha. The number of Woodcock being shot per annum has been increasing since the mid 1970s. However, this does not necessarily reflect an increasing Woodcock population. There may be, but as most Woodcock are shot on Pheasant shoots, the increased shooting of Pheasant may have also increased the bags of Woodcock.

Box 7.4

Direct counts for Grey Partridge.

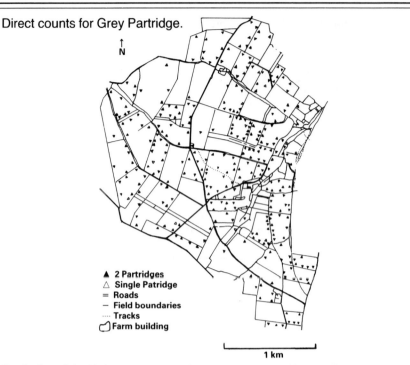

▲ 2 Partridges
△ Single Patridge
= Roads
— Field boundaries
···· Tracks
⌂ Farm building

1 km

By plotting all the birds seen on an early morning survey of farmland between mid March and late April, the number of Grey Partridge on the site can be counted and some idea of the population and their habitat preferences obtained (from Hudson and Rands 1988). For example, it is clear from this figure that most of the Partridges are associated with field margins, with relatively few seen in the centres of fields, and also few alongside the road running roughly north–south through the figure. Moreover, some fields appear to support more Partridges than others and the birds seem to be approximately equal distances apart in individual fields. Various hypotheses could be erected to explain the population level and its distribution within the survey area; these could then be tested on the ground. Similar surveys repeated after the breeding season would assess productivity and could be used to see if there is a shift in the positioning of the birds.

Specialised counting methods for breeding waders.

**Box
7.5**

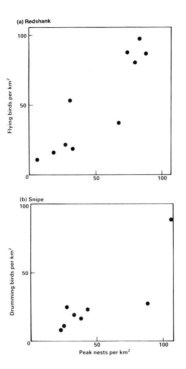

These graphs show the relationship between counts of birds per km^2 and peak numbers of nests per km^2 for (a) Redshank and (b) Snipe (Green 1985b). Each point represents the data from an individual study area. Peak nests per km^2 were calculated by locating nests by intensive rope-dragging experiments. Peak numbers of birds per km^2 were calculated somewhat differently for the two species. For Redshank, the mean density of birds seen per visit in April and May was used, and for Snipe, the mean density of drumming (aerial display) birds in April and May. At least three counts and usually five or six were made for each species. Fitted regression lines to these graphs (not shown) indicate that one Redshank counted in April and May represents approximately one nest at the peak period. For Snipe, one drumming bird is equivalent to about two nests at the peak time, although there are rather few data to substantiate this latter relationship.

Box 7.6

Survey sheet for the 1989 British Lapwing survey.

INSTRUCTIONS

AREA: Observers should count all nesting Lapwings in the tetrads (2 × 2-km squares) which have been selected at random by the National Organiser. All fields and other open habitats in each tetrad should be checked. Please return a card for every tetrad selected, even if apparently unsuitable or no Lapwings are found. **Nil returns are wanted.**

COUNTING: Lapwings are conspicuous when nesting and displaying. Territorial males, pairs or birds standing guard near nests are easily counted. On short vegetation incubating birds are easily seen and all fields should be carefully scanned for them. Counts of nesting pairs should be based on these points. Extensive open ground can be checked from roads, tracks and footpaths without disturbing nesting birds and this technique should be used whenever possible to ensure accurate counts. Even in apparently flat fields undulations can hide birds, so please ensure that all the field has been studied. Walking through a field (with permission), or a Crow flying over, will usually cause all the Lapwings to fly up, including any incubating birds. In this case the total seen may be halved for the number of pairs. **Please ask for access if it is necessary to enter private land away from public rights of way.**

TIMING: A complete count of each tetrad (which may need more than one visit to cover thoroughly) carried out in April is wanted. Nesting birds and nesting habitat being the aim, counts in southern counties may be best timed for the first half of the month, and in northern for the second half.

NOTES ON FILLING IN THE CARD:

1. Please record the total number of nesting Lapwings found in each habitat in each 1-km square in the appropriate box. If no Lapwings are found in a 1-km square put a zero in the totals box only. Any hatched broods seen should be recorded on a separate sheet.

2. Only record livestock if present during counts, noting C, S or H as appropriate, in the relevant box.

3. Farms may keep both cattle and sheep (and horses) and cattle and sheep may be grazed together. In these cases summarise the number of nesting Lapwings present with each type of stock in the relevant box as, e.g. 5C, 2S, 3 C+S etc.

4. Some autumn cereals are tramlined and some may have bare patches after the winter, which attract Lapwings. Where pairs nest in autumn cereals please note P (bare patches) or T (tramlines) if appropriate, e.g. 4T 5P 2 etc.

5. Record any pairs in oil-seed rape under Other habitats.

6. An enlarged copy of the $2\frac{1}{2}$ inch OS map of each tetrad is supplied with each card. These show field boundaries. On the map please record the number of pairs found nesting in each occupied field and the field's boundaries. The latter is important as many of these maps are rather out of date.

7. *Autumn and spring cereals* differ sharply in April. Autumn cereals are taller and/or bushier and usually darker green. Many spring cereal crops will still be at the single leaf stage, having just emerged.

8. *Tramlines* are the permanent wheel-tracks left across growing cereals.

9. *Other spring crops* are crops other than cereals planted in spring, e.g. potatoes, peas.

10. *Plough* is bare land still in furrow. Tilled land is bare land worked down fine but without an emergent crop. Please record bare land as one or the other, even if it is known to be planted.

11. *Permanent grass and ley* are often difficult to distinguish. Leys usually lack the varied plants, uneven structures and firm turf of permanent grass. Young leys left for forage may be very even and dense stands, resembling thickly-sown cereals rather than a grazed meadow.

12. *Don't know grass.* It is important to separate permanent grass and ley whenever possible, but when it is not please record the field under this heading.

13. *Rough grass* should be taken as poor grassland infested with rushes or weeds.

14. *Tetrads.* The lettering system for tetrads is illustrated right.

15. *Previous counts.* If you have any previous counts for this tetrad please enter year and number in the box provided or on a separate sheet.

E	J	P	U	Z
D	I	N	T	Y
C	H	M	S	X
B	G	L	R	W
A	F	K	Q	V

Box 7.6 cont.

SURVEY OF NESTING LAPWINGS IN ENGLAND AND WALES

British Trust for Ornithology

Please read instructions on reverse and fill in the map

County

10-km square no.

Tetrad letter

Main village or town

Village or farm

For office use

Observer: Name

Address

Was tetrad fully searched?

Dates of visits

NUMBER OF NESTING PAIRS OF LAPWINGS IN:

Habitat	grazed C S H*	ungrazed
Autumn sown cereals (notes 4,7,8)		
Spring sown cereals		
Sugar beet		
Other spring crops (note 9)		
Bare land (note 10) — Ploughed		
Bare land (note 10) — Tilled		
Stubble		
Ley grass (note 11)	grazed C S H*	ungrazed
Perm-anent grass (note 11)	grazed C S H*	ungrazed
Moor, rough grass (note 13)	grazed C S H*	ungrazed
'Don't know' grass (note 12)	grazed C S H*	ungrazed
Saltmarsh		
Dunes		
Gravel pits		
Waste ground		
Reservoirs		
Sewage farms		
Airfields		
Other (specify) (note 5)		
TOTAL		

TOTAL

Any previous counts? (note 15).

*C = cattle. S = sheep. H = horses (notes 2, 3)

Please return by 30th June to Regional Organiser or to BTO, Beech Grove, Tring, Herts. HP23 5NR.

In this survey sheet, data on the number of birds recorded, their activity and habitat they were found in are requested. This enables both population and habitat studies to be completed. Results of this census are presented in Shrubb and Lack (1991).

Box 7.7

Specialised mapping of territories for Tawny Owls.

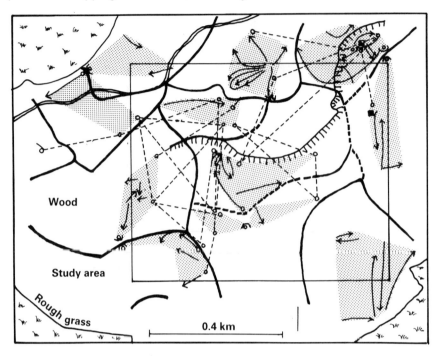

The figure presents Tawny Owl territories in Wytham Great Wood in 1954 and 1955, as determined by observation (from Southern and Lowe 1968). The large square is the detailed study area. Small circles connected by dashed lines show positions of birds giving territorial challenges ('hooting') simultaneously and thus separated. Brackets show territorial boundary disputes. The shading indicates the minimum area of a territory from the observation of territorial challenges. Black lines with arrows show observed movements and direction. Heavy black lines show the best fit of territories determined from observations and prey items marked with rings and recovered from pellets. The interrupted parts of these heavy lines show where changes of boundaries occurred before the year 1955/56. The black squares represent nest-boxes used for breeding and the inner triangle is the trapping grid.

This method is similar to the CBC but only territorial disputes can be used to separate territories. Interpretation of the results can produce density estimates for this species and can allow the position of territories over time to be mapped, the mortality and longevity to be investigated and the habitat relationships of the birds to be understood.

Reproduced with permission of the Institute of Terrestrial Ecology.

3. Herons

Methods for counting breeding populations of an obvious tree-nesting species and a secretive species are presented below.

OBVIOUS SPECIES

The *Grey Heron* breeds colonially in trees. Care must be taken when visiting colonies as disturbance may cause the birds to abandon their nests.

The counting unit is the Apparently Occupied Nest. These should be counted in the late incubation or early nestling period (late April in Britain). Nest occupancy can be ascertained using the following criteria: (1) eggs in nest, (2) eggshells beneath nest, (3) young seen or heard, (4) adults sitting, (5) fresh nesting material found, (6) droppings on or below the nest. It may be difficult to distinguish between individual nests and in such cases the number of Apparently Occupied Nests must be estimated. Detailed counts in a sample of heronries in Scotland have shown that these methods record about 70% of the pairs using a site (Marquiss 1989). It may therefore be possible to produce correction factors for heronry counts.

In large study areas, all heronries, or a random sample of them, may be counted to produce an accurate picture of changes in the breeding population. Colonies naturally increase and then wane, with new smaller colonies being set up. Hence, in long-term studies of a regional heron population the new colonies must be located, otherwise counts will be biased towards showing a population decline.

SECRETIVE SPECIES

The *Bittern*, inhabits extensive reedbeds and is extremely secretive during the breeding season, hence its breeding population is very difficult to count.

Traditional methods involve counting vocalisations (booms) over the breeding period, February until June or July. Booming is of greatest intensity in the early morning and dusk, hence survey visits are recommended at these times. At least three survey visits should be undertaken to assess population levels.

The 'booming' male is the counting unit. During survey visits, positions of 'booming' birds are marked on a map of the site. Attempts should be made to count the whole site simultaneously to avoid problems when birds move. The number of territories is interpreted as is described in Chapter 3.

It may be possible to obtain a more accurate count of Bittern populations by electronically analysing the boom-patterns of all birds in a study area which are known to be specific to a single male (RSPB, unpublished).

4. Wildfowl

Populations of larger and more obvious species (swans and geese) are counted using the look-see method, and populations of smaller and more secretive species by a variety of specialised methods. Examples of each are presented below.

OBVIOUS SPECIES

For the *Mute Swan* the counting unit is the territorial bird with a nest. Counts should be made when the territories are well-defined and the nests conspicuous (April or May in Britain).

Males vigorously guard their breeding territories from all potential intruders. Proof of breeding is the location of an active nest or brood of cygnets. Potential bias in counts results from poor coverage of waterbodies, and because non-breeding males will sometimes set up and defend territories even though they are not paired.

The Mute Swan population for the whole of Britain has been estimated by counting territorial pairs in 10-km square counting units in April and May (Ogilvie 1986). Box 7.2 shows the standardised census form used in a 1990 Mute Swan census.

SECRETIVE SPECIES

Breeding populations of many *Anas* and *Aythya* duck species are difficult to count as they nest in dense vegetation and often move their broods to other areas as soon as they hatch. Three counting methods are commonly used.

(1) Counts of nesting females. The counting unit is the female with nest. The most accurate method of locating nests involves rigorous searches in suitable habitat (e.g. Hill 1984a,b). However, foot-searches for nests are extremely labour intensive and may result in nest desertion; consequently they are rarely undertaken.

In America, line transect methods have been developed to count the number of females nesting in prairie grasslands. One method involves driving two jeeps with a 50 m cable chain between them through the grassland along a transect of known length. The number of females flushed per unit area is counted and used to assess densities (Duebbert and Lokemoen 1976). A second method utilises light aircraft to fly transects and count ducks flushed from their nests (Bellrose 1976).

(2) Counts of off-duty males. The counting unit is the male duck. The number of males in small groups are counted just after the females have started to incubate their eggs and thus have become highly secretive. Groups of males counted should comprise fewer than five birds to exclude flocks of non-breeding or late wintering birds.

Different species of duck should be counted at different times during the spring. For example, in a climatically average year in southern Britain, Mallard start nesting in March/April, Shoveler in late April/early May, Teal in April, Pochard in April and Tufted Duck in May. However, these timings can be shifted forwards or backwards according to the weather in a particular year, hence the best counting periods for most ducks will vary by year.

Pöysä (1984) recommends a series of counts spread throughout the potential breeding period to identify the optimal census period for each duck species in a particular year. This time is recognised as when the maximum number of males have flocked together, after the wintering flocks have dispersed and before any post-breeding flocks return.

(3) Counts of duck broods. The counting unit is the female duck with young brood. Counts can be made by direct observation of a site over a designated period, or by flushing broods onto the open water by walking the banks with dogs (Rumble and Flake 1982). Flush counts are generally more successful and quicker than observations, except on larger or more vegetated waterbodies.

5. Raptors

Counting breeding and non-breeding raptors poses special problems as most are found at low densities and often use specialised nesting habitat in remote and inaccessible areas.

BREEDING RAPTORS

Look-see methods are commonly used to assess breeding populations of raptors, e.g. for Golden Eagle (Watson *et al.* 1989). During such studies it is particularly important that equal time is spent studying each site in detail.

The counting unit is the Apparently Occupied Nest-site, or Breeding Territory. Proof of occupancy by a pair would be (1) seeing two birds together, (2) finding moulted feathers or droppings, or preferably (3) finding a nest containing eggs or young, or seeing adults carrying food or hearing the begging calls of young birds.

OBVIOUS BREEDING RAPTORS

For the *Buzzard* the counting unit is the soaring bird. These should be counted when they are on their breeding territories and when the ground has warmed sufficiently to allow soaring.

The British breeding population has recently been assessed by counting birds within randomly located 10-km squares throughout Britain in the early spring (Taylor *et al.* 1988). Population estimates were produced by assuming one soaring bird was equivalent to one pair.

SECRETIVE BREEDING RAPTORS

For the *Sparrowhawk* the counting unit is the nest-cluster, where clusters include nests from previous years. Counts are made in the early spring once the birds have set up their nesting territories.

Recommended methods of locating nest-clusters are the observation of adult birds in or near a woodland, followed by searches of the woodland floor for droppings, feathers from plucked prey, pellets, or the nests themselves (Newton 1986).

In Britain, nest-clusters are regularly spaced over extensive areas, being between 0.5 and 2.0 km apart depending on their geographical location (Newton *et al.* 1986). Once one nest has been located a useful method to locate the next nest is to draw a circle of appropriate diameter on a map of the region and then visit any woodlands intersected to check for nesting Sparrowhawks.

DISPERSED NON-BREEDING RAPTORS

In North America line transects are commonly used to assess non-breeding raptor populations over extensive areas. In general, these surveys rely on cars for transport and roads to provide the transect routes. The method involves driving slowly (17–40 km per hour) and counting all birds that one or two observers detect, usually within a specified distance (0.4–1.6 km) on each side of the road, and on calm and clear days (Fuller and Mosher 1981: Chapter 4). This method is valuable for the obvious soaring species, but cannot detect the more secretive forest-dwelling birds. Variability in the detectability of raptors, and consequently errors in density estimates derived from these counts, are fully discussed in Millsap and LeFranc (1988).

ROOSTING RAPTORS

Populations of some raptors can be assessed by counting birds at their roosting sites. In the recent count of wintering Hen Harriers in Britain (Clarke and Watson 1990) questionnaires were widely distributed to local conservationists and amateur ornithologists, requesting data on known or suspected roosts. A co-ordinated programme of watches of all roosts was then organised and the number of birds present in mid-winter (January) was assessed.

MIGRATING RAPTORS

Migrating raptors (and other large species) can be counted at bottlenecks along their migration routes. For example, raptors on migration from Africa to Europe can be counted at The Bosporus (Turkey), Eilat on the Red Sea and the Straits of Gibraltar. The most complete migration-route counts are made over the entire migration period, but 80–90% of the birds can be recorded over 2–3-week windows, the dates of which are known for the more important routes. To count the birds, teams of observers (ideally one to three counters, one identification checker and one transcriber) are positioned 6–8 km apart across the flyway. The exact location of the counting site should give (1) the best available view of the centre of movement and (2) the best possible views of the birds to make identification as easy as possible; in general these are areas of higher ground. If there is a wide migration front and observers can be spaced the recommended 6–8 km apart there is little likelihood of double counting, but with closer spacings these risks are increased. Teams count the number of gliding birds (not those circling in thermals) passing per hour. It is useful for one observer to count to the north, one to the south, and one overhead. Counts should be carried out and recorded in hourly units, starting on the same hour to facilitate comparison of results between sites and years.

6. Gamebirds

There are a variety of methods available to count these species. Examples are grouped for convenience into upland and lowland species.

UPLAND SPECIES

There are three methods commonly used for *Red Grouse*.

(1) Direct observation of territorial encounters. The counting unit is the territorial male. These can be counted from late autumn to spring, as males are on territory, frequently display and have territorial disputes (Hudson and Rands 1988). Males can be counted from a vantage point, or whilst walking or driving transects across suitable habitat.

(2) Pointing dogs which locate all birds. The counting unit is the individual Red Grouse, both territorial and non-territorial. Maximum detectability by dogs is in the autumn (October to December) and spring (March and April) (Hudson 1986). The total number of birds in the study area can be assessed using this method, but it is difficult to define the number of birds holding breeding territories.

(3) Shooting bag records. Useful information on population fluctuations and trends can be derived from bag records from shooting estates (Box 7.3: Tapper 1989). In Britain, such data are available for the past century, but are influenced by several factors apart from population level, hence their main use is in long-term studies of population dynamics (e.g. Barnes 1987).

Three methods are also used for *Capercaillie*.

(1) Counts of males and females at lekking sites. The counting unit is the male and/or female at the lekking site. These should be counted from a suitable vantage point at dawn in the early spring (mid to late April in Britain; Moss and Oswald 1985). Lekking sites can be identified by systematically searching suitable areas, looking for tracks in snow, or field-checking information provided from local sources (Rolstad and Wegge 1987). If all lekking sites have been located, an accurate assessment of the breeding population can be made. If it is unknown whether all leks have been recorded, then the mean of the maximum number of birds (cocks and hens) seen at lekking sites may be used to produce an index of the breeding population.

(2) Counts of females with broods. The counting unit is the female with chicks. Counts should be undertaken during July when females are moving around with their broods (Moss and Oswald 1985). Females are counted by two persons with pointing dogs quartering the study area or walking transects 20 m apart; the dogs should locate all the birds and the presence of chicks can then be ascertained and mapped. If the number of females with

broods has been assessed from a number of sample plots in a larger site the density of breeding females can be extrapolated for the whole site.

(3) Drives with beaters. The counting unit is the individual bird. These should be counted on drives undertaken from late October to early November, when most birds have taken to the trees (Lindén and Rajala 1981; Moss and Oswald 1985). Several people (beaters) arranged in lines 20 m apart, walk slowly through suitable habitat to flush the birds.

LOWLAND SPECIES

Two methods are used for *Grey Partridge.*

(1) Spring count. The counting unit is the individual bird. These are counted in March, when the birds have paired and before the vegetation gets too high. Counts are made in the 2 hours after dawn and the 2 hours before sunset, when the birds should be feeding in the open. Calm and dry conditions are preferred (Potts 1986). Birds are counted by an observer using binoculars from within a vehicle. Using this method, up to 200 ha of farmland can be covered in about 2.5 hours. Birds show up well in newly planted crops which can be counted from the field margins, but the vehicle has to crisscross well-grown grasslands. On each visit maps showing the location of birds should be produced (Box 7.4). The maximum number of pairs plotted on the survey maps is used to produce a population estimate for the study area.

(2) Post-breeding stubble count. The counting unit is the individual bird, or pairs with chicks. These are counted in the first 2 hours after sunrise during August. The post-breeding population can be estimated from the peak counts, and the breeding success can be quantified from the number of young birds observed and the proportion of pairs with no chicks.

7. Rails and Crakes

Many of these species are very secretive during the breeding season and most species are associated with wetland habitats where they may be exceedingly difficult to view. Methods for counting obvious and a highly secretive species are presented.

OBVIOUS SPECIES

For *Coot* and *Moorhen* the counting unit is the breeding territory. Counts should be made early in the breeding season (late March or early April in Britain). Several visits are recommended during this period to obtain the best population estimate.

All evidence of breeding birds along the vegetation/water interface is recorded on maps of the study site on each visit. Coots display and fight more than Moorhens, making mapping of their territories easier.

Population estimates are produced using methods outlined in Chapter 3. Criteria used to define a territory are: (1) lone bird, (2) pair (two birds together), (3) territorial dispute (= two pairs) and (4) calls of birds unseen (Koskimies and Väisänen 1991).

SECRETIVE SPECIES

Methods of counting *Corncrake* have been developed in northern Scotland and Ireland (Cadbury 1980; Stowe and Hudson 1988).

The counting unit is the breeding territory. Calling birds (presumed to be mainly males) are counted on several visits to the study area between 23.00 and 03.00 hours BST in late May, June or even early July. The location of calling birds is plotted on a map of the study area on each visit. Play-backs of the craking call may be used to stimulate the birds into calling.

The number of breeding territories is assessed using methods outlined in Chapter 3. A territory is regarded as being occupied if the bird is heard calling for more than 5 days (Stowe and Hudson 1988).

8. Waders

Two general sets of methodologies have been developed for counting breeding waders in Britain: field-by-field counts (Fuller *et al.* 1983, 1986; Barrett and Barrett 1984; Green 1985a,b; BTO 1989), and transect methods (Reed and Fuller 1983; Reed *et al.* 1984, 1985; Avery 1989).

FIELD-BY-FIELD METHOD

This method is most suitable for fine-grained habitats. It was used in the Breeding Waders of Wet Meadows Survey in Britain (BTO 1989). Large-scale maps of study-sites showing site boundaries and numbered individual fields were issued to all counters and three counting visits were recommended:

Visit 1—between 15th April and 30th April;
Visit 2—between 1st May and 21st May;
Visit 3—between 22nd May and 18th June.

If possible, sites were to be visited before 12.00 hours BST, cold, wet or windy weather was to be avoided and successive visits were to be at least 1 week apart. All fields were to be walked to come within 100 m of all points, and fields were scanned 200–400 m ahead to check for displaying larger waders.

TRANSECT METHOD

This method is most suitable for open habitats. Reed and Fuller (1983) have recommended counting birds along transects in a series (between two and four) of visits in May and June. Transects are located between 50 and 200 m apart depending on the 'anticipated' density of breeding birds and all birds recorded are marked on maps. Counting is conducted between 09.00 and 17.00 hours BST as this avoids the confusing periods of maximal bird activity in the early morning and evening. Avery (1989) modified the method to cover larger areas by using only one observer to traverse transects 200 m apart, and by having only two visits during the breeding season (May and June).

Methods used to assess population levels for individual species of wader are presented below.

SECRETIVE GRASSLAND SPECIES

The counting unit for *Snipe* is the displaying (drumming) male. These should be counted on at least three occasions during the display period (April and May in Britain), and within 3 hours of dawn or dusk (Green 1985a). Rope-dragging experiments, which discover all Snipe nests in a study field, have indicated that the true nesting population can be calculated by doubling the mean of April/May counts of drumming birds (Box 7.5).

The counting unit for *Redshank* is the flying bird showing alarm. These are best counted when the birds have young (late May to early June in Britain), between 09.00 and 17.00 hours BST. At this time both parents will mob an observer traversing a field or walking set transects, and both should be counted individually. Several visits spread over a couple of weeks are recommended to increase the accuracy of the results. Detailed work by Green (1985b) indicates that the mean number of Redshank recorded during the recommended period equates to the maximum number of nests present (Box 7.5). However, the population can be estimated by halving the number of flying birds recorded on a single survey visit.

OBVIOUS GRASSLAND SPECIES

The counting unit for *Lapwing* is the incubating bird. Counts are recommended when the birds are sitting on eggs (late March to late April in Britain) because later juveniles, finished and failed breeders flock and confuse the count (Reed and Fuller 1983; Barrett and Barrett 1984). Incubating birds are located by carefully scanning the study area. Several counts should be made over the recommended period, between 09.00 and 12.00 hours BST when activity is most stable. Detailed experimental work has concluded that the maximum of a series of counts made during the period when most pairs are incubating gives a good estimate of the number of birds breeding (Green 1985b).

A recent survey of breeding Lapwings in Britain counted birds in a randomised selection of 10-km squares throughout the country using volunteers and the form shown in Box 7.6.

BARE-GROUND NESTING SPECIES

For *Ringed Plover* and *Little Ringed Plover* the counting unit is the territorial bird which is best counted when it is incubating (Parrinder 1989; Prater 1989).

The recommended counting technique is to scan 50–100 m ahead and count all visible birds, then walk on rapidly and repeat the process. However, because the birds are inconspicuous, careful scanning is important. Little attempt is generally made to prove breeding by finding nests or broods as this causes disturbance and is very time-consuming. There are also problems with keeping track of all the birds during the count; individuals may undertake fast pursuit flights over large areas which can lead to overestimation.

LARGELY NOCTURNAL SPECIES

For *Stone Curlew* the counting unit is the incubating bird. These are counted at dusk and during the night. Birds are located by playing tapes of their call from a slowly moving vehicle. If the taped call is within 500 m of an incubating bird it will answer and can thus be counted.

Woodcock is a difficult bird to count accurately as it spends the day in woodland and feeds on fields during the night. There are three counting methods.

(1) Counts of displaying males. The counting unit is the displaying 'roding' male. The counting period is throughout the breeding season (April to end June in Britain), with the maximum activity in Britain occurring towards the end of May (Hirons 1980). Populations can be only roughly estimated because some males rode more than others and the birds do not occupy discrete areas.

(2) Drives with beaters. The counting unit is the individual bird. These are counted after being flushed by teams of beaters and dogs. Drives should be undertaken during the winter in the day-time. As drives allow all birds to be counted they permit densities to be calculated.

(3) Nocturnal feeding counts. The counting unit is the individual bird. These are counted as they fly to or from nocturnal feeding areas at dusk or dawn. Counts can be undertaken throughout the year and allow indices of population level to be produced.

UPLAND SPECIES

For *Golden Plover* the counting unit is the calling bird. These should be counted once the chicks have hatched (generally by mid June in Britain) because at least one parent remains on guard near the chicks and calls loudly and persistently if anyone approaches. Before this time Golden Plover are secretive. Birds should be counted by walking transects 100 m apart through areas of suitable habitat. Population estimates are produced by assuming each calling bird equals one pair. Transect recording in association with intensive nest searches has suggested that the method locates over 80% of the breeding pairs (Yalden and Yalden 1989).

Dunlin is an extremely difficult species to count accurately as it breeds semi-colonially, has small territories, does not move far to mob intruders, and is very inconspicuous (Reed and Fuller 1983; Reed *et al.* 1984). The counting unit is the individual bird. These should be counted along transects during the day (between 09.00 and 17.00 hours BST in Britain). Greatest detectability in Scotland has been found in June, as once the birds are incubating they are more difficult to flush and are even more subject to underestimation.

For *Curlew* and *Whimbrel* the counting unit is the displaying bird. These are counted on three visits to the study area in the late incubation/early fledging period (late May/early June in Britain). Birds are counted whilst the observer walks transects 200–400 m apart, or from the edge of a field. All birds showing signs of being on territory are marked onto 1 : 2500 scale maps in the field.

For Whimbrel, population estimates are produced from the identification of breeding territories based on criteria such as (1) alarm calls most intense, (2) both birds of a pair calling actively overhead or undertaking distraction displays, (3) one of the pair adopting a characteristic secretive 'creeping' run, (4) both birds of a pair alighting near observer and calling in a highly agitated state (Richardson 1990).

9. Owls and Nightjars

These birds are all mostly nocturnal or crepuscular, and modified methods of counting their breeding numbers are required.

TERRITORY MAPPING METHODS

The counting unit for the *Tawny Owl* is the territorial male. Counts are made in late autumn (October–December in Britain) when territorial activity is most intense (Southern and Lowe 1968; Mead 1987). All records of 'hooting' and other calls should be plotted on maps during biweekly visits to the study area. Territories are identified from boundary disputes between males and clusters of records (Mead 1987; Box 7.7).

For the *Nightjar* the counting unit is the calling (churring) male. These should be counted at dusk on calm days (Beaufort force 0–4) throughout the breeding period (May to July in Britain). The number of churring males is assessed as the maximum number of birds that call from separate locations at dusk and for the next half-hour. Separate locations are generally defined as sites over 500 m apart with calling less than 30 seconds apart (Cadbury 1981; Gribble 1983). Large numbers of people are required to record individual calling locations on a large site and thus produce accurate population estimates.

TRANSECT METHODS

In North America, populations of owls are counted by recording the number of birds responding to calls played from tape-recorders. Calls of several species (starting with the smallest) are played at fixed distances (0.4–1.6 km) along transects defined by roads and return calls listened for. The calls are repeated several times to allow birds to reply. All returned calls are noted and are used to produce population estimates (Fuller and Mosher 1981).

LOOK-SEE METHODS

For the *Barn Owl* the counting unit is a bird on its territory. The breeding population of Britain has been assessed by widely distributing questionnaires on the bird, allied with publicity campaigns, local knowledge and interviews with farmers (Bunn *et al.* 1982; Shawyer 1987). Only the proof of a bird on its territory has been required to define a breeding pair.

In more detailed population studies Barn Owls have been counted by rigorous searches for nests. The procedure is to mark down on detailed maps of the survey area the location of all buildings, groups of trees, etc. that may provide suitable nest-sites. All potentially suitable areas are then rigorously cold-searched for nesting birds. In simple areas with few potential nesting sites this method is practical, in other areas it may prove too time consuming.

10. Corvids

Various methods are available to count the colonial and non-colonial members of this group. Examples of methods used in Britain are provided below.

COLONIAL SPECIES

The counting unit for the *Rook* is the Apparently Occupied Nest. These should be counted in the spring (April) before the leaves are fully unfurled to aid location of the tree-based colonies (Sage and Vernon 1978). Apparent occupancy is assessed from criteria such as (1) birds bringing nesting material, (2) birds sitting on nest, (3) birds landing on nest.

NON-COLONIAL SPECIES

The counting unit for the *Crow* is the Apparently Occupied Nest. Nesting birds should be counted in the spring (April) before the leaves have fully unfurled. Nests are located singly in trees, hence all suitable habitat in a study area must be carefully scanned for potential nests. Occupancy is determined by (1) birds bringing nesting material, (2) black feathers and droppings present beneath nest, (3) birds observed flying to or from the nest, (4) birds observed on the nest.

The counting unit for the *Raven* is the Occupied Territory. This can be determined early in the year (January/February in Britain) by assessing the number of birds at potential nesting sites. Ravens are conspicuous and their nests are built in easily recognisable sites, are bulky and persist for several seasons. In the study of Marquiss *et al.* (1978), information on the number of Ravens in a study area was built up from a combination of local knowledge and active searching for nests. In general pairs are regularly distributed in an area of similar habitat, so by plotting the distribution of known pairs on a map, gaps become apparent which can be checked on the ground. Over a period of years the exact number of Ravens in an area can be determined.

11. Other Passerines

Breeding populations of the majority of passerines are counted using mapping methods, point counts or transects (see Chapters 3–5). However, there are groups of species, such as those nesting in extensive wetland vegetation, where the habitat is so difficult to census that no reliable counting methods have been developed. British examples of such species are Reed Warbler, Sedge Warbler and Bearded Tit. There are many other species in the world where no useful method of counting their population has been developed.

Summary and points to consider

Look-see population surveys involve studies of areas of habitat thought to be suitable for the study bird, usually during the breeding season. It is important to be familiar with the study area and the ecology and behaviour of the bird concerned. Similar levels of effort should be expended in counting each potential site to avoid counts reflecting effort rather more than number of birds.

Modified mapping methods can be used to count individual species which have interactive encounters with others of the same species.

Modified transect methods can be used to count certain individual species, especially larger or more obvious ones.

Drives with beaters are a useful way to count larger ground-birds which can be scared into flight. If the area beaten is known, then the density of the bird can be calculated.

Shooting bag records provide an index of the population level of hunted species. Bag records are, however, also influenced by factors other then population such as shooting pressure.

There are many species that are not easily counted by the standard methods of territory mapping, transects and point counts, and are also not amenable to the more specialised methods presented in this chapter (e.g. rainforest canopy species, many wetland passerines). At the present time assessing their populations remains extremely difficult.

8

Counting Colonial Nesting and Flocking Birds

Introduction

Counting colonially breeding seabirds and non-breeding flocks of birds presents special problems which must be addressed if accurate counts are to result.

In a seabird colony these problems include the difficulties of assessing the proportion of breeding and non-breeding birds, locating and counting breeding colonies on remote and rugged coastal sites, evaluating the proportion of birds that have left the nest to obtain food, and defining the effects of harsh weather on numbers of birds at the colony.

In flocks the problems include the limitations of binoculars and telescopes, the variability in the ability of observers to identify species of birds in flocks or to estimate numbers of birds within a flock, especially when several species of different sizes are intermingled.

Despite these problems, however, procedures have been developed to count birds in breeding colonies and in flocks. These are outlined below.

Seabird breeding colonies

The various stages in counting seabird breeding colonies are presented below.

1. Description of study area

The region to be surveyed should be visited and the position of all seabird colonies and other breeding areas marked on a base-map at 1 : 10 000 scale (Box 8.1).

If colonies are spread along an extensive length of cliff, or one colony cannot be viewed from one site, or if breeding density is high, then the study area should be divided into counting sections dependent on the availability of suitable vantage points.

2. Description of breeding colony

A colony is defined as a concentration of breeding birds separated from others by an area of cliff, sea, or open space. If in doubt it is usually best to sub-divide a colony, so long as this can be done unambiguously.

For each colony the following should be recorded (after Seabird Group/ NCC 1988; Lloyd *et al.* 1991).

(1) Colony name. Names taken from base map and the same as on national 1 : 50 000 map of study area.

(2) Location, e.g. north side of Firth of Forth, near Crail. Six figure grid references from the base-map for the start and finish of the cliff section, or the approximate centre of a colony on flatter ground should also be given.

(3) Status, e.g. Statutory Nature Reserve, Private Nature Reserve, private landowner (specify owner if possible).

(4) Description. Details of cliff height and orientation, shore slope, rock type (geological map), vegetation cover (type and amount), or main habitats if a flatter site or an island. If possible sketch these details in the field and take Polaroid photographs as a permanent record (writing date and details of colony on back of photograph). Location of counting positions and direction of view should also be marked (Box 8.1). It is important that the boundaries of the colony or sample plots are shown in relation to the main features of the region, streams, gullies, etc. so that they can be located exactly in the future.

(5) Access. How to get to the site, boathandler's name and address, landowner's name and address, etc.

(6) History. Counting history if known, with bibliography where possible.

(7) Counting problems. Indicate approximately what percentage of the colony can be counted from land, how much can be seen from the sea and any particular counting problems, e.g. birds nesting in caves, counted whilst looking up, broad ledges hiding birds, restricted view of colony, disturbance of colony by observer.

(8) Other notes. Any relevant information on the colony, e.g. site of annual population monitoring etc.

(9) Bibliography. Any details of books, scientific papers, reports, etc. that mention the colony.

3. Selection of counting method

The aims of the count and the species present will largely determine the methods used. A rough estimation of breeding numbers of some seabird species over a large geographical region can be achieved using rapid and relatively crude methods such as aerial surveys. However, full population estimates for a defined study area usually require more detailed and time-consuming methods. For example, counting populations of ground-nesting seabirds (e.g. Herring Gull) may involve counting nests in quadrats located in a colony, and counting populations of cliff-nesting seabirds (e.g. Common Guillemot) may involve several counts of individual birds in well defined areas of cliff at particular stages in the breeding cycle.

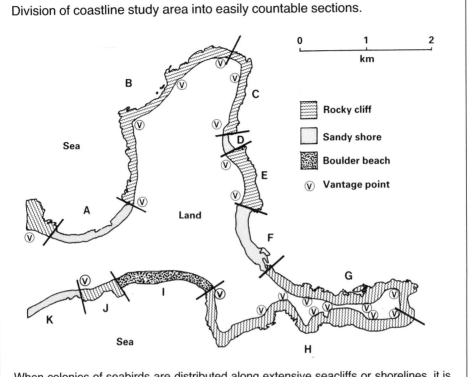

Division of coastline study area into easily countable sections.

Box 8.1

When colonies of seabirds are distributed along extensive seacliffs or shorelines, it is necessary to divide the cliff or shore into easily countable sections. These are best defined by the features of the area (vertical cliff, boulders, sandy beach, etc.), the availability of suitable vantage points from which the birds can be counted, and because each section can be counted easily in one go. It is important that all sections and vantage points are marked on the base-map of the study area, and the results of the counts are presented according to the various sections. It is also worth considering counting the sections in the same sequence, or randomly, to minimise bias caused by colony attendance altering over the counting period.

4. Counting the birds

To obtain the most accurate counts at cliff-nesting colonies the position of the observer is important. Ideally, observers should be at the same level, or slightly above, the birds and should be looking directly at the colony (Box 8.2). If this preferred position cannot be obtained, the observer is forced to count the birds from available locations. As cliffs are dangerous places observer safety should be considered an overriding priority in the selection of the counting position.

Box 8.2

Positioning of observer for counting seabirds breeding on cliffs.

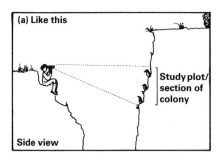

(a) Like this

Study plot/ section of colony

Side view

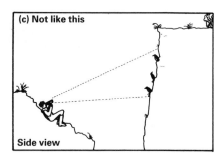

(c) Not like this

Side view

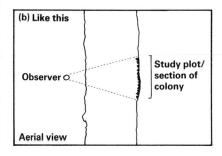

(b) Like this

Observer ○

Study plot/ section of colony

Aerial view

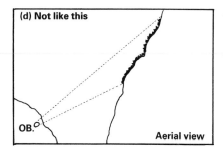

(d) Not like this

OB.○

Aerial view

Correct and incorrect positionings of observer for viewing and counting study plots or whole colonies of breeding seabirds are shown (from Birkhead and Nettleship 1980). (a) Side view—observer should be slightly above breeding birds; (b) aerial view—observer should be directly opposite the study plot. In examples (c) and (d) poor positions for counting birds are presented. However, observer safety should always be considered the number one priority and if the ideal positions for counting birds cannot safely be obtained, then a compromise should be reached.

Reproduced with permission of the Minister of Supply and Services Canada, 1991.

Census form for counting breeding seabirds.

Box 8.3

SEABIRD COLONY REGISTER

Data Sheet

FOR OFFICE USE

Name: _____ Year: _____

Give address on back of sheet if
different from Colony Register Form

Colony Name: _____

Notes: Use back of sheet

County or District: _____

FILL IN HERE

SPECIES			DATES OF COUNTS	ACCURATE COUNT	RANGE OF ESTIMATE min.	max.	Unit	Method	Br. Status
Fulmar	022	ı ı							ı
Manx shearwater	046	ı ı							ı
Storm petrel	052	ı ı							ı
Leach's petrel	055	ı ı							ı
Gannet	071	ı ı							ı
Cormorant	072	ı ı							ı
Shag	080	ı ı							ı
Arctic skua	567	ı ı							ı
Great skua	569	ı ı							ı
Black-headed gull	582	ı ı							ı
Common gull	590	ı ı							ı
Lesser black-back	591	ı ı							ı
Herring gull	592	ı ı							ı
Great black-back	600	ı ı							ı
Kittiwake	602	ı ı							ı
Sandwich tern	611	ı ı							ı
Roseate tern	614	ı ı							ı
Common tern	615	ı ı							ı
Arctic tern	616	ı ı							ı
Little tern	624	ı ı							ı
Guillemot	634	ı ı							ı
Razorbill	636	ı ı							ı
Black guillemot	638	ı ı							ı
Puffin	654	ı ı							ı

FILL IN HERE

UNIT
1 = Individual bird on land
2 = Apparently occupied nest
3 = Apparently occupied territory

COUNTING METHOD
1 = From land 4 = From photo
2 = From sea 5 = From land and sea
3 = From air 6 = Other, give details in Notes.

BREEDING STATUS
01 = Bird in habitat
02 = Singing in habitat
03 = Pair in habitat
04 = Territory
05 = Display
06 = Nest site
07 = Anxious parent
08 = Incubation
09 = Nest building
10 = Distraction
11 = Used nest
12 = Fledged young
13 = Occupied nest
14 = Food for young
15 = Nest + eggs
16 = Nest + young

(Continued)

<table>
<tr><td>

Box 8.3 *cont.*

</td><td>

In Britain a standardised form has been produced for the purpose of counting breeding seabirds (from Seabird Group/NCC 1988; Lloyd *et al.* 1991).

Numbers of birds counted and estimated are entered in separate columns, so that the two parts of the census when added together give an approximate total for the colony. The following codes define the counting units.

1 = Individual birds on land, excluding any on non-breeding ledges or loafing areas; 2 = apparently occupied nest-sites; 3 = apparently occupied breeding territories; 4 = other, give details in notes.

The following codes define the counting methods.

1 = From land; 2 = from sea; 3 = from air; 4 = from photo; 5 = from land and sea; 6 = other, give details in notes.

The following codes define the level of certainty that the species breeds in the colony.

01 = Bird in suitable nesting habitat during the breeding season; 02 = bird singing in suitable nesting habitat during the breeding season e.g. petrels; 03 = pair of birds seen in suitable nesting habitat during the breeding season; 04 = bird seen defending territory, two records at least 1 week apart; 05 = courtship displays recorded; 06 = nest-site found; 07 = agitated/anxious parents seen; 08 = bird seen incubating; 09 = bird seen building a nest; 10 = distraction display recorded; 11 = used nest found, e.g. broken eggshells, droppings, food remains, etc.; 12 = fledged young present—not used for species that may have travelled some distance e.g. petrels; 13 = occupied nests, contents unknown; 14 = food seen being brought to young; 15 = nest with eggs found; 16 = nest with chicks found.

</td></tr>
</table>

Seasonal variations in numbers of nests and clutches at breeding colonies.

**Box
8.4**

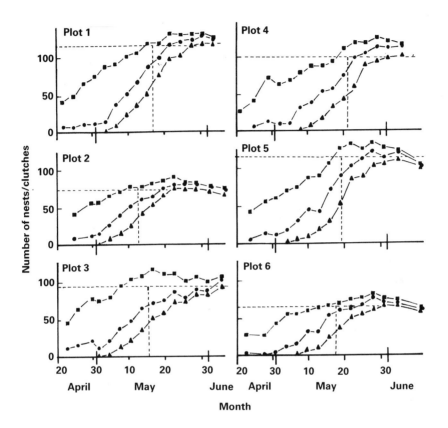

This figure shows changes in the total number of nests (■), complete nests (including those containing eggs) (●) and clutches (▲) of Larus gulls in six plots on the Isle of May in Scotland between April and June 1983 (from Wanless and Harris 1984). The horizontal dashed line indicates the number of pairs breeding in the plot (taken to be the highest count of clutches plus nests that had lost clutches during the 12 days prior to that count). The vertical dashed line shows the median date of laying of the first egg. This figure indicates that the population of breeding gulls reached a peak in late May and began to fall in June. Nest-counts should be made during the period of maximum nests (late May) and counts in future years should be made at the same time.

Box 8.5

Randomly distributed quadrats for counting ground-nesting seabirds.

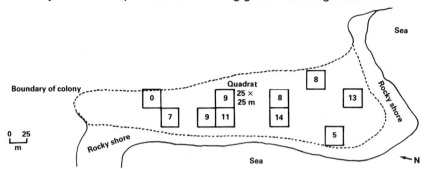

In this example the extent of the colony was first mapped. Then the colony area was overlain on a grid of 25 × 25 m, and quadrat locations of this size that fell wholly within the colony were chosen by random numbers generated on a calculator. Numbers of birds were counted in each of the ten quadrats and the population of the whole colony could thus be calculated from the following data.

Number of nests in quadrats = 84.

Area sampled = 6250 m².

Nesting area = 39 550 m².

$$\text{Number of nests in the colony} = \left(\frac{\text{Nesting area}}{\text{Area sampled}}\right) \times \text{Nests in quadrats} = 532.$$

This method would be wrong if the colony density differs between the edge and the centre, since edge quadrats are excluded from the sample. A more correct method would be to lay out a grid that covers the entire colony and then count randomly located squares, whether or not they fall wholly within the colony.

Seasonal variation of breeding seabird numbers at breeding colonies.

Box
8.6

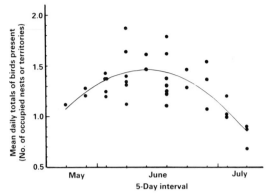

This figure presents the nest attendance of Arctic Terns at colonies on Orkney and Shetland in 1980 (from Bullock and Gomersall 1981). By plotting the variation in the number of birds over the season these authors have demonstrated that the best census period for this species in northern Scotland is mid June as numbers are increasing up until this time, and begin to fall afterwards. With other species, or in other regions, the form of the graph may differ and indicate a different ideal counting period. The regression line shown on the graph fits the regression equation $y = 0.8496 + 0.2433x - 0.0242x^2$ ($P < 0.025$).

Comparison of observed and expected numbers of Common and Roseate Terns in the Azores using a ratio of three flushed birds equalling two breeding pairs (del Nevo unpublished data).

Box
8.7

Colony and species	Flush count	Expected pairs	Known pairs	Difference	%Difference
(1) Common Tern	70	47	45	2	4.4
(2) Common Tern	14	9	8+	1	12.5
(3) Common Tern	28	19	18	1	5.6
(4) Common Tern	120	80	83	3	3.6
(5) Common Tern	190	127	120	7	5.6
(6) Common Tern	126	84	85	1	1.2
(7) Roseate Tern	32	21	21	0	0

Box
8.8

Statistical variation of breeding bird numbers at cliff-breeding colonies over the breeding season.

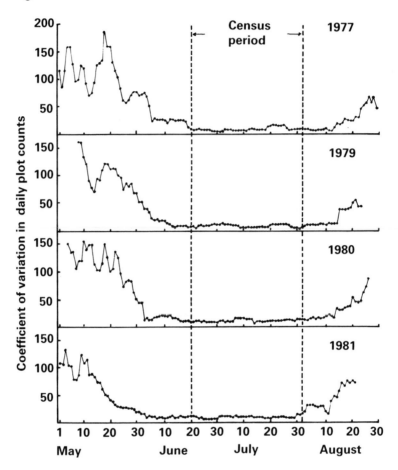

This figure presents seasonal changes in the coefficient of variation (CV) among daily plot counts of Guillemots over 4 years on the Semidi islands, Alaska (from Hatch and Hatch 1989). Each plotted value is the coefficient calculated for a 7-day interval with the indicated date at its midpoint. This technique shows that variation among daily counts of plots is lowest between 20 June and 30 July showing that counts should be made at this time when they would be quite reliable. For other birds, or in other regions, the form of the graph may vary and similar studies might be necessary to elucidate the ideal period for counting the birds.

Transects for counting populations of burrow-nesting seabirds.

Box
8.9

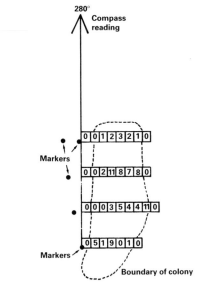

In this example, transects of 5 × 5 m quadrats are positioned continuously across an area of seabird nesting burrows. The estimate of the number of burrows within the colony was produced using the following data (from Nettleship 1976).

Number of quadrats in colony = 98.

Number of quadrats sampled in colony = 24.

Number of *active* burrows in quadrats sampled = 88.

Average number of *active* burrows per quadrat = 3.67.

Number of *active* burrows in colony = 3.67 × 98 = 359.

Reproduced with permission of the Minister of Supply and Services Canada, 1991.

| Box 8.10 | Methods of estimating numbers of birds in flocks (modified from Howes 1987). |

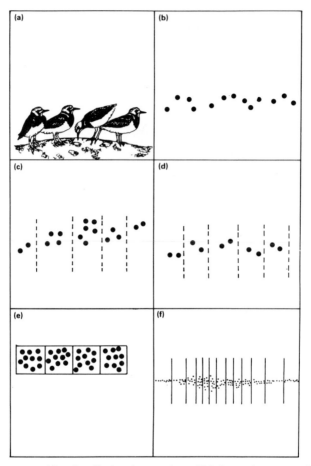

(a) In small roosts and feeding flocks, the number of birds can be counted directly.

(b) For small flying flocks of even density, the birds can be counted individually (1, 2, 3, 4, 5, etc.) to produce an accurate total. If a suitable landmark is present it can be used to help to count the birds.

(c) In unevenly distributed flocks with small groups of varying size, each group of birds should be rapidly counted and added together.

(d) For larger numbers of birds in evenly distributed flocks the birds should be counted in multiples e.g. 2, 4, 6, 8, or 3, 6, 9, 12, etc. Again if landmarks are present they can be used to help divide the flocks in order to count them more accurately.

(e) For densely packed flocks in flight or at a roost, the birds should be counted in estimated blocks. The size of the blocks used (10, 100, 1000, etc.) varies according to the size of the flock. The largest flocks of 10 000 birds or more present the biggest counting problems with even the block method giving a very rough estimate of numbers.

(f) Flying flocks often bunch in the centre. In this case it is important that the blocks are closer together in the centre of the flock than towards the edges, but in practice this may be difficult to achieve.

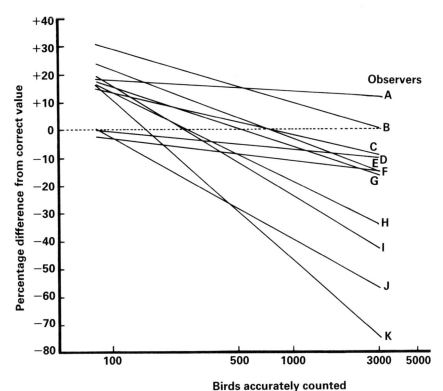

Trends in the accuracy of different observers at counting flocks (from Prater 1979). **Box 8.11**

This figure shows that overall there was a tendency for observers A–K to over-estimate slightly (10–20%) the number of birds in smaller flocks (100–400), but to under-estimate increasingly numbers in larger flocks. For example, in flocks of 3000 birds estimates were consistently low by around 25%. However, individual observers vary in their ability to estimate numbers of birds. For example, observer A consistently produces estimates between 10 and 20% above the true number, but the accuracy of the estimate hardly varies with increasing numbers of birds. In comparison, observer K estimates 10–20% more birds in flocks of 100 individuals, but this rapidly changes to an estimate around 70% below the true figure when counting flocks of 3000 individuals. This indicates that the accuracy of different observers is highly variable and needs to be checked in any formal counting programme.

5. Use of standardised census forms

As with other census methods, standardisation of approach and recording is vitally important when counting seabirds. Standard forms have been produced in many countries and for many purposes. The standard form of the British Seabird Colony Register is shown in Box 8.3. This illustrates the type of information which should be collected during a seabird count.

All such census forms stress the importance of recording the year, month, date, and phase of the breeding cycle (pre-laying, incubation, chick-rearing) when the counts were made. Standardisation of data enables them to be compared with those from other areas and between years.

Counting methods for various groups of seabirds

Detailed descriptions of counting methods for various seabirds are provided below. The bulk of the information has been taken from a small number of publications (Nettleship 1976; Birkhead and Nettleship 1980; Evans 1980, 1986; Seabird Group/NCC 1988; Lloyd *et al*. 1991).

1. Gulls

The recommended counting unit for *Kittiwake* is the Apparently Occupied Nest-site. This is defined as a substantial or well-constructed nest capable of holding two or three eggs and occupied by at least one bird on or within touching distance of the nest (Seabird Group/NCC 1988). An active nest is usually obvious, owing to a covering of white faeces. Apparently Occupied Nest-sites are counted during the late incubation to early nestling period (Nettleship 1976; Heubeck *et al*. 1986; Harris 1987), generally early to mid June in Britain. When a whole colony is being counted, the count is made easier if the cliff is divided into sections and these are counted separately.

Some non-breeding birds build nests but do not lay. These nests are generally less well built and less obvious than active nests. If possible they should be excluded from counts; also guano-stained loafing sites should be identified and similarly excluded. The final counts should ideally be the mean of at least three counts of the same section of the colony on the same day. Repeat counts of particularly dense colonies on different days are a valuable way of checking the results, and if the same observer undertakes all counts their variation can be calculated. However, if time is short, a single count of the site in June provides a good estimation of numbers as the variation in colony attendance at this time is low.

Nest-counts made from photographs of the colony are also useful, but should be treated with caution as the status of some sites is difficult to interpret.

Several methods are available for counting *Larus gulls* (*e.g. Herring Gull, Black-headed Gull, Lesser Black-backed Gull, Great Black-backed Gull and Common Gull*).

(1) Full nest-counts. For this method the counting units are Apparently Occupied Nest-sites. These are defined as the summed number of occupied and unoccupied nests that appear to have been used during the present breeding season (Box 8.4). This caveat is applied because gulls have precocial young and some nests may have fledged young when others still contain young or have not yet had eggs laid in them. If a good vantage point is available and the colony contains fewer than c. 200 pairs then all Apparently Occupied Nest-sites can be counted directly during the mid incubation period, late May to June in Britain.

In the British Isles it is recommended that full nest-counts are made between 09.00 and 16.00 hours BST as colony attendance is most stable during this period. It is also recommended that counts are not made during periods of heavy rain, fog or high winds as these are believed to affect the accuracy of the count (Wanless and Harris 1984).

(2) Nest-estimate using transects. In less easily viewed, or larger, colonies the number of Apparently Occupied Nest-sites, as defined on the criteria presented above, may be estimated using transects through the colony.

The first stage is to map the extent of the colony either from a ground survey, or more rapidly from aerial photographs, and mark the boundaries on a base-map.

The second stage is to define transects through the colony to obtain a representative sample of the population, and where possible mark these with coloured string. Each transect is walked by an observer and the number of Apparently Occupied Nest-sites 0.5–1 m either side of the transect line are counted and marked by tags or paint to avoid double counting. If the area of the colony and the area of the transect are known, then the number of breeding pairs in the colony can be calculated (see Chapter 4).

(3) Nest-estimate using quadrats. It is also possible to use quadrats to sample the gull population and thus derive a population estimate. Quadrats should be between 5×5 m and 20×20 m depending on the density of the colony. They can be placed at equal distances along a transect, or randomly within the colony (Box 8.5). A good methodological example of the use of quadrats to count ground-nesting seabirds is provided by Thompson and Rothery (1991).

When using transects or quadrats it is important to minimise the time spent in the colony. Thirty minutes is the maximum period observers should remain in a colony, and if the birds are disturbed within the first few minutes the observers will have to withdraw until they settle down again. Prolonged disturbance may lead to egg loss (e.g. predation), or chick loss (e.g. wandering from their nests and being either attacked or killed).

Two other less precise but much more rapid methods are available to count nesting gulls.

(4) Flushing counts. In this method the counting unit is the flying bird. All the gulls are flushed from their nests and rapidly counted, and the number divided by two to give an estimate of the number of pairs. The method is primarily useful in isolated locations and especially on small islands (Haila and Kuuesla 1982; Hanssen 1982). The technique involves the observer(s)

startling the birds by raising and lowering their arms whilst standing on the highest point in the area, or by making a loud noise. This causes the gulls to leave their nests and fly around where they can be counted using methods outlined later (see Terns). For this method it is especially important that disturbance is kept to a minimum length of time as predators may steal nestlings and eggs, and the birds may abandon their nests.

(5) Aerial counts. By flying over a colony a rough estimate of the breeding population can be made by counting all sitting birds and assuming they have nests. More accurate counts can be obtained from aerial photographs taken whilst flying over a colony. These can fix the colony boundaries precisely and allow the nesting birds to be counted back on land. Disadvantages of an aerial survey are the expense, poor level of counting accuracy, high level of disturbance which may cause birds to desert their nests, and difficulties of identifying breeding and non-breeding birds.

2. Terns

Terns are prone to moving their colonies between years, hence it is extremely important to search the study area thoroughly to determine the distribution of colonies before the counting commences. Three methods have been developed to count the numbers of breeding birds at colonies of Common, Little, Arctic, Sandwich and Roseate Terns.

(1) Direct counts. For this method the counting unit is the Apparently Occupied Nest-site, defined as those birds sitting tight and apparently incubating eggs or brooding chicks. Ideally these Apparently Occupied Nest-sites are counted from a position where the whole of the colony can be viewed. However, problems occur when the whole colony cannot be viewed, and where both members of the pair sit slightly apart and both are counted as incubating. Nevertheless, this method produces accurate results if used with care and is most useful in smaller colonies.

(2) Flushing counts. For this method the counting unit is the flying bird. The method was developed to count Arctic Terns in the Scottish islands of Shetland and Orkney (Bullock and Gomersall 1981), and has subsequently been applied in counting all terns in Ireland (Whilde 1985) and Common and Roseate Terns in the Azores.

The procedure is to flush all birds present at a colony into the air using a loud noise (e.g. fog horn) and then count the birds several times whilst they are in the air. By averaging these counts the mean number of birds at the colony can be calculated.

Bullock and Gomersall (1981) showed that the timing of these counts was important if accurate results were to be obtained. Counts made throughout the incubation and post-incubation period, starting in late May (first egg-laying) and continuing until mid July (first chicks fledging), showed a peak in the number of birds in mid June (mid incubation to early nestling) (Box 8.6). It was concluded that this is the best counting period for this species in Shetland and Orkney. Bullock and Gomersall also counted birds throughout the day and showed that colony attendance was stable between 08.00 and

22.00 hours BST. Hence it was concluded that in this area counts could be made between these hours.

In other regions similar validation counts may need to be carried out to assess diurnal and seasonal variation in colony attendance.

Flushing counts have been related to the true number of nesting pairs by the calculation of a nest-attendance index (Bullock and Gomersall 1981). This was calculated by counting terns using both direct counts of Apparently Occupied Nest-sites and flushing counts at a small number of 'calibration colonies' every 5 days over the breeding season at 2-hourly intervals between 08.00 and 22.00 hours BST. It was discovered that three flying birds were equivalent to two breeding pairs. By using this calibration figure an estimate of the total breeding population was made for the whole study area.

Close correspondence has also been demonstrated between flushing counts and direct nest counts for colonies of Common and Roseate Tern in the Azores (Box 8.7).

(3) Transect counts. For this method the counting unit is the Apparently Occupied Nest-site as defined above. Firstly, the total extent of the colony is assessed on the ground or from aerial photographs. Then transects passing through the colony are used to sample, and hence assess, the breeding population. Ideally more than one transect should cross the colony and around ten 5 × 5 m quadrats positioned along this transect are used to count the number of Apparently Occupied Nest-sites. This method may cause particular disturbance and hence can be used only in the less dense colonies.

Whatever method is being used to count terns, the observer should never remain in the colony for longer than 20 minutes as terns are highly sensitive to disturbance and may desert the site, have their eggs taken by predators, or be trampled by inexperienced observers. Indeed on many occasions the terns will become so agitated in the first few minutes that the observer will have to retreat and allow the birds to calm down.

3. Auks

The counting unit for the *Common Guillemot* is an individual at the breeding colony (Birkhead and Nettleship 1980; Evans 1980). These are best counted during the middle egg-laying to middle chick-rearing period, as colony attendance is most stable during that period (Box 8.8).

Counts also vary with the time of day (see Chapter 2). To minimise these diurnal effects, counts in Britain are made between 06.30 and 16.00 hours BST (10.00–13.00 hours BST is preferred). However, other studies have noted a different diurnal attendance pattern (Evans 1986; del Nevo 1990) so it may be necessary to obtain colony-specific diurnal attendance patterns before detailed counting begins.

Counts should be made on 5–10 separate days over the counting period (Box 8.8) and results averaged to even out between-day differences in attendance. If possible the number of birds nesting in crevices and cracks should also be estimated.

Calculating the number of breeding pairs from the counts of individuals is difficult because the breeding density and number of non-breeding birds present at the colony varies over the breeding season (Harris 1988). Correction factors, or k values, have been developed to calculate numbers of breeding pairs at a Guillemot colony. These k values are calculated from designated 'control' ledges where the number of birds on nests (N_p) and the mean number of individuals (N_i) are carefully assessed from 10 separate counts during the census period, i.e. between the last egg being laid and the first chick hatched. The k value is then calculated as $k = N_p/N_i$ (Birkhead and Nettleship 1980).

Counts of individuals should be multiplied by the mean k value for the colony to obtain an estimate of the number of breeding pairs. Since attendance patterns vary between colonies, derived k values may be colony-specific and should not be used at different colonies (but see Harris 1988).

Guillemots can also be counted from photographs of the colony. However, such counts are generally unreliable because the birds are often not obvious on photographs: when viewed from the front they blend in with guano-covered ledges and from the back with rock and shadows. Moreover, the number of non-breeders cannot be deduced from photographs (Birkhead and Nettleship 1980).

For *Razorbill*, individuals form the counting unit and these should be counted in the incubation to early nestling period. Counts should be made between 07.00 and 15.00 BST in temperate regions, but between 05.00 and 15.00 hours BST in the Arctic (Birkhead and Nettleship 1980; Evans 1986). As the diurnal attendance pattern may vary between sites this should be checked before a counting programme is initiated.

It is usually impossible to calculate correction values (k values) for Razorbills because they rarely form discrete colonies, the birds often nest in scree slopes or boulders in inaccessible locations, and it is difficult to assess the number of breeders and non-breeders.

The counting unit for *Black Guillemot* is the adult-plumaged individual. Counting the breeding population directly is difficult because nests are generally out of view in cracks and gullies and the species generally nests at low density. As a consequence, the recommended counting method is to walk along the top of rocky shores and low cliffs, or drive a boat along the base of cliffs in the early morning (05.00–09.00 hours BST) in the pre-breeding period (April to early May), and attempt to flush all birds out onto the sea. All adult-plumaged birds on the sea within 200–300 m of the shore should then be counted, and all immature-plumaged birds noted separately. These counts are best repeated 3–5 times on separate days to produce an average count, and preferably should be made only when there are calm sea conditions and winds less than Beaufort force 4 as higher winds make counting extremely difficult and colony attendance is altered. Counting in the early morning is important as the birds fly out to sea later on in the day (Birkhead and Nettleship 1980; Ewins 1985). As the proportion of non-breeding birds is very variable and the birds are difficult to see, k values are not useful for this species.

For *Puffin* the counting unit is the Apparently Occupied Nest-site, which in burrow-nesting colonies is defined as a burrow sufficiently deep for a Puffin to nest in, showing signs of being occupied (Apparently Occupied Burrow) (Harris and Murray 1981; Harris 1983; James and Robertson 1985; Harris and Rothery 1988). Occupancy can be determined by the presence of fresh diggings or droppings in the burrow entrance, and during the nestling period by the presence of broken eggshells, discarded fish, a fishy smell or directly with an optical fibrescope. Rabbit burrows are easy to separate from those used by Puffins using these methods, but Manx Shearwater burrows cannot be reliably separated. In mixed colonies, careful observation early in the morning (before 07.00 hours BST), when Puffins bring fish to the burrow, may be required to determine those burrows occupied by Puffins and those by Manx Shearwaters.

The density of occupied Puffin burrows is best assessed using a sampling procedure such as randomly located quadrats or transects (Box 8.9). In the study of Harris and Rothery (1988) 56 circular quadrats of 30 m^2 in each area were randomly positioned throughout an extensive Puffin colony; random numbers were used to define quadrat-locations on a map and these were marked with stakes in the field. By knowing the area of the quadrats and that of the colony the sample results could be used to extrapolate the number of occupied burrows in the whole colony.

Probably all that can be achieved on cliff-nesting Puffin colonies is to count individual birds on land and the sea close to shore to produce a crude index of the breeding population (Nettleship 1976). These counts are best carried out just before dusk when the birds come in to the colony. It is important to record land and sea counts separately from burrow counts, and to note the time of day (BST), weather conditions and length of count.

4. Other species

The counting unit for *Fulmar* is the Apparently Occupied Nest-site, defined as an individual sitting tightly on a reasonably horizontal area large enough to hold an egg (Nettleship 1976). Two birds on a site, apparently paired, count as one site. It is difficult to determine the number of breeding pairs because prospecting birds may occupy a site for several years before producing an egg and may be confused with breeding pairs.

Counts of Apparently Occupied Nest-sites should be made in the late incubation to early nestling period when the colony attendance is greatest, usually in late June or early July in northern Europe (Dunnet *et al.* 1979). Several counts are generally made over a period of 3–7 days to reduce problems with colony attendance varying between days, and a mean number of Apparently Occupied Nest-sites calculated. Counts should also be made in the middle of the day (12.00–13.00 hours BST, maximum 09.00–16.00 BST) as attendance is highest at that time.

The counting unit for *Manx Shearwater* is the Apparently Occupied Nest-site, defined as a burrow of sufficient depth to support a Manx Shearwater and showing signs of occupation (Apparently Occupied Burrow). Occupancy can be determined from the presence of droppings and scrapings over the breeding period, a fishy smell in the burrow, directly with an optical fibrescope, or by recording the number of responses to a tape of Manx Shearwater calls played down the hole at night (James and Robertson 1985). Counts of Manx Shearwater colonies should be made shortly after the completion of egg-laying (June in Britain). Usually the limits of the colony are defined on a base map and then samples of the number of occupied burrows using randomly located quadrats or transects are used to extrapolate the total colony population from this sample. Circular sampling plots are recommended. They can be created using ropes tied to a central stake, for example a rope of 1.78 m will produce a circular quadrat of 10 m^2, 2.52 m of rope will produce a quadrat of 20 m^2, and 3.09 m of rope a quadrat of 30 m^2. The smaller quadrats should be used in high density colonies and larger quadrats in increasingly lower density sites. About 30 quadrats should be recorded in each colony to facilitate comparisons between years and sites (Wormell 1976).

On Skomer Island in Britain attempts have also been made to assess Manx Shearwater numbers using capture–recapture ringing methods (see Chapter 6).

The counting unit for *Northern Gannet* is the Apparently Occupied Nest-site, defined as those sites where an individual is sitting tightly on a reasonably horizontal area large enough to hold an egg. Two individuals sitting next to each other, and apparently paired, count as one site, as do single sitting individuals. All sites should be counted irrespective of whether or not any nesting material is present.

The count is made in the late incubation to mid nestling period, usually June. Two counting methods have been developed.

(1) Direct counts. Counts of Apparently Occupied Nest-sites can be made directly from boats or from the land.

(2) Counts from photographs. Photographs are an easy method to assess the status of a colony as expanding colonies always increase in area, and declining ones decrease. Gannet numbers are also relatively easy to count from photographs as Apparently Occupied Nest-sites are large and regularly spaced (Harris and Lloyd 1977). Slides taken from a boat or the air can be projected onto a screen and then individual Apparently Occupied Nest-sites blocked out as they are counted. Tests of observers counting birds from photographs indicate that observer error in the counts is usually less than 15% and with experienced observers may be less than 10% (Murray and Wanless 1986).

The counting unit for *Great* and *Arctic Skuas* is the Apparently Occupied Territory, as assessed in the mid incubation to mid nestling period, approximately early June in northern Scotland (Everett 1982; Furness 1982; Meek *et al.* 1983). Two counting methods exist.

(1) Direct counts. The procedure is to select a suitable vantage point and scan the study area using binoculars. Beware of counting paired birds standing apart as two territory holders, and of overlooking birds that blend against the background. The counts should be repeated on three separate

days spread over the counting period, with all Apparently Occupied Territories marked on a map.

(2) Transect counts. In this method, the procedure is to walk a transect through the study area. Pairs that show annoyance or feign injury when an observer passes through their nesting area can be counted as Apparently Occupied Territories and marked on a map. Possible errors in this counting method are (i) both members of a pair may be counted as two separate territory holders, (ii) birds may be overlooked, or in the air, or out of the territory, when the count is conducted, (iii) difficulties may occur in subdividing large colonies into separate territories, (iv) counts late in the season may be higher than those earlier as a few young birds may establish territories.

The counting unit for *Shag* is the Apparently Occupied Nest-site, defined as substantial or well-constructed nests, occupied by at least one individual (Evans 1986); bear in mind that the species has a long breeding season and there may be some nests that have fledged young when others have eggs, and that some sites are used by more than one pair (Harris and Forbes 1987). Recent research on the Isle of May in northeast Scotland shows that the number of nests increases to a peak in early June and then declines. In this study the peak count of Apparently Occupied Nest-sites was 89% of the true annual total number of nests (Harris and Forbes 1987), suggesting that a single count of Apparently Occupied Nest-sites in early June produces a reliable index of the breeding population, at least in this part of Scotland.

Monitoring breeding seabird populations

As well as complete counts of the breeding birds in a study area, it is also important to be able to monitor seabird populations annually to assess if populations over large areas are changing. In the British Isles there has been much progress towards developing a unified seabird monitoring programme over the past 15 years (Stowe 1982; Evans 1986; Mudge 1988; Harris 1989; Lloyd *et al.* 1991). This information is held as a part of the Seabird Colony Register run by the Joint Nature Conservation Committee, the Institute of Terrestrial Ecology and the Royal Society for the Protection of Birds.

In this scheme, the monitoring plots are well defined areas, usually a colony or a group of birds within a colony, where annual counts of the breeding seabirds take place. As a general rule, and depending on the species being studied, these monitoring sections should include 50–100 pairs of cliff-nesting seabirds.

The position of the monitoring site within the colony is important. The plots should aim to provide a representative sample of the colony. Ideally plots would be randomly located throughout the colony, but in practice randomly located plots may be impossible to count. At the present time most monitoring plots have been pragmatically selected for their ease of counting and believed representativeness, i.e. plots encompass most of the variation in the colony, including some edge, but they avoid areas where birds are particularly densely packed, where they are extremely difficult to count.

Details of the methodology used to establish a monitoring plot for Common Guillemot is presented below (after Harris 1989).

Select several study plots (e.g. five) dispersed through the colony where there are 50–100 nests which can be viewed from the same level or from above. Ideally these would be randomly distributed throughout the colony but observer safety and difficulties of viewing in most colonies dictate that at most sites their location will be chosen non-randomly. This can be done by dividing a colony into four or five approximately equally sized sections and picking one or two plots within each section, trying not to bias plots towards the centre or edge of the colony.

Take photographs of the monitoring plots from a good vantage position when the birds are incubating or brooding small young (June in Britain). Large-scale photographs (20 × 20 cm) are essential for the first year, but in subsequent years the outline of the colony, important features and location of study plots can be traced from the original photograph. Tape overlays onto the original photograph so it can be annotated in the field.

View the area from where the photographs were taken, at approximately the same time of year. Plot the positions of (1) birds with an egg, (2) birds with a chick, (3) birds apparently incubating, (4) pairs regularly attending a site that appears capable of supporting an egg (bearing in mind that some eggs are laid on unsuitable sites).

Make several visits until satisfied that most of the occupied sites have been located. Record any chicks without an adult in attendance. Number the active sites. To assess breeding success the contents of active sites should be noted every 1–2 days. Any young leaving when aged 15 days or more and/or are well feathered can be considered as having been raised successfully.

If assessing breeding success, present the results as x young fledged from y active (i.e. 1–3 above) and z inactive (i.e. 4 above) sites as found on the dates of the first checks.

Make notes if you have any reason to suspect the season, or the results, may have been atypical.

Follow the same areas each year.

Similar methods can be used to monitor populations and breeding success of other species of cliff-nesting seabirds.

Flocking birds

Many species of birds form flocks for roosting, feeding and protection. These flocks are very difficult to count using standard methods (Chapters 3–5) and specialised counting methods have been developed.

The first stage in the counting procedure is to mark the boundaries of the areas to be counted on a map. This allows the same areas to be counted in the future to assess population changes.

Wherever possible try to position the observer with the sun behind, and in some cover. This enables birds to be more easily viewed and identified and not scared away. For smaller sites it is desirable to count the whole site from a single viewing position to minimise disturbance and the possibility of birds being counted twice. However, this will not be possible on larger sites and the site will have to be divided into counting sectors. The boundaries of the sectors will depend on the habitat being surveyed, e.g. for counts of roosting waders on the coastline, the sector will be a length of coast that can be covered by a counter within 2 hours of a high spring tide, preferably with obvious landmarks at either end.

1. General counting methods

Counts of roosting, flying and feeding flocks are made using two main methods.

(1) Direct counts. If the congregation is no more than a few hundred birds a suitable vantage point should be located and all the individual birds counted directly using binoculars or a telescope. This is easy with large birds close up, but becomes progressively more difficult with large numbers, smaller birds and greater distance. Eventually estimation methods have to be used (Box 8.10).

(2) Estimation counts. If there are large numbers of birds, especially if the flock is mobile, specialised estimation procedures should be used to count them. Firstly, a good vantage point is located (e.g. where birds can be seen entering or leaving the roosting site, or where the feeding habitat is over-looked), then the birds in the flock are counted in mentally-divided groups of e.g. 5, 10, 20, 50, 100, 500, 1000 depending on the total number of birds in the flock and the size of the birds (Box 8.10). Landmarks can be used to help break up large flocks into more manageable sections. If possible, counts should be repeated several times and another observer's opinion obtained on the number of birds before a final count is recorded. Problems arise when snap estimates have to be made of a rapidly moving and large flock. Counts should be expressed in terms of the species forming the flock (may be mixed), the total number of birds, and an estimation of the numbers of the various species if the flock is mixed, or the proportion of each species.

2. Example of counting roosting birds

The example chosen here is that of roosting waders. On the coast wading birds are forced to retreat to higher ground during high tides. On such occasions most of the wader population on, for example, an estuary will be concentrated at a few high-tide roost sites where they can be counted. The British Birds of Estuaries Enquiry (Prater 1981; Kirby 1987, 1990), coordinated by the BTO and WWT, collects data on wader populations at a large number of British estuaries, allowing total population levels, population changes and seasonal patterns of movement to be assessed.

Counts of waders at their roost sites are made 2 hours either side of high tide on the highest spring tide of the month that lies on a Sunday in hours of daylight, preferably close to the middle of the month. Sunday is chosen because most of the counters are volunteers and this is the best day for them. Counting effort is focused on the winter period (September to March) when movements tend to be least and the numbers of birds highest. The counting area is divided into sectors which are counted every year. The monthly counts are coordinated so that all roosting sites within a single estuary are counted as near simultaneously as possible, by a team of counters if the site is large. The maximum peak count for the winter period at a particular estuary is published annually (e.g. Kirby 1987; Salmon 1989). To obtain comparative data for peaks between years at least five counts are required over a winter period, since it is almost certain that some of them will have to be discarded (because of incomplete coverage, poor visibility, disturbance, etc.) To obtain a reliable index to enable comparison with the first year of counting it is also important that there is a continuity of counters and that everyone knows the area well.

Wader roosts are usually traditional and it is important that sufficient time is expended to locate them prior to any counting. To locate roosts, all suitable habitats such as saltmarsh, shingle, beaches and spits should be visited on a rising tide when birds are beginning to congregate. Birds will also congregate on short pasture, recently tilled or rolled arable fields or recently harvested fields up to 1 km inland. Where possible these sites should also be checked and counted where practical.

In small roosts (a few hundred waders) individual birds can usually be counted from a suitable vantage point at high tide when all the birds are in the roost. Larger roosts, and those comprised of small species, are more difficult to count accurately, and considerable care must be taken when arriving at totals. One, or a combination, of the following techniques is usually successful.

(1) Count all the birds as they fly from their feeding grounds to roost sites, repeating counts where possible. Counts should start at least 2 hours before high water.

(2) Count the stationary birds whilst they are roosting at high tide, repeating the counts several times. This is the best method as long as the birds are not too tightly packed, as is often the case for small species such as Dunlin and Knot.

(3) Count birds on the ebbing tide when they are leaving the roost, repeating the counts where possible. This method works particularly well for those species that disperse quickly from the roost to start feeding e.g. Dunlin and Redshank.

At some roosts a combination of all three methods will be needed to produce accurate totals, and with roosts of smaller species which are tightly packed the most reliable estimates will be obtained when the whole flock is in flight and can be counted using methods outlined in Box 8.10. Estimates are generally recorded in parentheses e.g. (3400). Generally speaking counting takes place over a 3–4-hour period (2 hours before and 2 hours after high tide).

Some species of wader are difficult to count using the high-tide-roost technique. These include species such as Purple Sandpiper which roost on rocky shores, and Lapwing, Golden Plover and Snipe which mainly roost inland. Other species, such as Knot, may be highly mobile at high tide and close liaison between counters of adjacent sectors and simultaneous counts are necessary to produce accurate totals.

3. Examples of counting feeding and flying flocks

The example described here is that of wintering flocks of waders and wildfowl. In the British Isles, feeding and flying flocks of wildfowl and waders are often counted as a part of the Birds of Estuaries Enquiry. Also, counts of wildfowl are made at inland waters as part of the National Wildfowl Counts, coordinated by The Wildfowl and Wetlands Trust (Owen *et al.* 1980).

British National Wildfowl Counts contribute to the International Water-fowl Census (Rose 1990), whose counting dates are defined internationally as the middle of January in the northern hemisphere and the middle of July in the southern hemisphere. Methods used to count flocks are outlined in Box 8.10.

In wildfowl counts the study area is normally counted in defined sectors, preferably ones that are used from year to year. For example in low-tide wader counts the intertidal area is divided up, either using natural features, or by positioning canes or poles in the mud at predetermined intervals. These sectors are counted on a cyclical basis with the same pattern of visits being undertaken on each count. All birds feeding and moving within the areas are counted every half-hour. Where possible the observer counts from the first appearance of mud to low water, or from low water until all the mud is covered.

4. Errors in estimating size of flocks

The numbers of birds estimated in a flock may be incorrect for many reasons: flocks may contain a very large number of birds, they may have a rapid swirling movement, there may be an interchange of birds between different flocks, the species within a single flock may be of considerably different sizes, some of the birds within the flock may be hidden at any one time, and there may also be problems of poor visibility or with the limitations of binoculars and telescopes.

A few studies have been made on the errors involved in counting flocks of birds (Prater 1979; Rapold *et al.* 1985).

Prater (1979) attempted to quantify observer error when estimating the size of flocks. Observers were asked to assess the number of birds on a large photograph which had been accurately counted using a binocular micro-scope. This method found that although individual observers differ in their ability to estimate the number of birds in a flock, the level of error generally varied with the number of birds being estimated (Box 8.11). In Prater's study estimates had to be made within 30 seconds, whereas in the field counts may sometimes be made over much longer periods and repeated several times before a final number is written down, hence repeated field counts may be more accurate than counts from photographs. However, snap estimates may also have to be made in the field for large numbers of small species briefly disturbed from a roost; these may be much less accurate than the counts obtained from photographs.

Prater (1979) has also shown that experience appears to affect the accuracy of counts produced from photographs of wader flocks. In general the least experienced counters produced the least accurate results.

The detailed studies of Rapold *et al.* (1985) also indicate large observer errors in the estimation of flying flocks, especially of smaller species. More details can be found in their paper.

Summary and points to consider

1. Breeding seabird colonies

These must be located, described and the birds counted using appropriate methods. Methods vary according to species but are well standardised.
Gulls and terns are counted as Apparently Occupied Nest-sites directly, along transects, within quadrats, or after flushing.

Auks are counted as individual birds during the peak nesting period. Counts of individuals can be used to produce estimates of pairs.

Burrow-nesting species are counted using sampling procedures.

For monitoring purposes, groups of birds in colonies should be counted in sections, or within well defined study plots.

2. Flocking birds

Roosting birds are often counted as they fly in or out of their roost. In smaller roosts, where the birds are visible, birds can be counted directly.

Feeding and flying flocks are best counted by dividing a site into sectors before counting the birds. Flocked birds are either counted individually, in small groups, or as blocks, depending on the size of the flock.

The errors in counting flocks may be considerable and tend to increase with the size of the flock.

9

Distribution Studies

Introduction

A species' distribution can be expressed simply as its presence or absence, or by some measure of abundance, across a set of sample units. The sample units may be on a regular grid such as employed in most bird atlas studies, or a random point within a habitat, at which bird data are collected.

There are essentially three types of distribution of animal species and communities: random, regular or aggregated (Box 9.1). Birds rarely show a random distribution because this implies they are distributed independently of features on the ground and independently of the presence of other birds. Further, the resources that birds exploit are rarely randomly distributed. For example, songbirds defending breeding territories in a woodland are more likely to be distributed regularly, if within-wood habitat patchiness is taken into account, whereas Sand Martin colonies tend to be aggregated.

The description of the distribution depends on the scale at which the birds are observed which in turn depends on the objectives of the study and the species concerned. Some birds use whole continents during their life-time (e.g. Arctic Tern) whilst others are so sedentary that their whole life may be spent in one particular woodland. The breeding distribution of a territorial woodland bird at the scale of the whole of England, for example, will be aggregated because of the distribution of woods. At the woodland level (or 2 × 2-km scale) however, they may be distributed more or less regularly because of territorial behaviour e.g. for Sparrowhawk (Box 9.1b). The dispersion patterns of several forest breeding species change as the size of area analysed changes (Wiens 1989). It is important to understand that answers to questions on bird distribution at one scale will not be provided by studying them at a different scale.

A knowledge of the distribution of a species is important because (1) the distribution can be related to land-use, (2) many of the conservation needs of a particular species or community can be identified by investigating habitat preferences which may manifest themselves through patterns of distribution (see Chapter 10), (3) the relative value of sites of conservation importance and vulnerability can be assessed with respect to their bird fauna, (4) information valuable to environmental impact assessments is provided, (5) baseline information is generated against which future changes can be assessed. Changes to populations may be more obvious from range changes exemplified by a distribution map than from counts (i.e. measures of abundance) taken at the centre of the range.

Distribution studies have been used to identify local, regional, national and international ranges of birds, habitat determinants of bird numbers, the effects of weather, arrival times of migrants, the extent of partial migration, patterns of influx by irruptive species, conservation importance of a particular species, threats to a site and a site's value to conservation.

There are three basic types of distribution study, each of which can be conducted at different scales of detail.

(1) Atlas studies: the bird distribution is considered at the international, national, regional or local scale, i.e. on a 'large scale'. Generally atlas studies are presence/absence of bird species, or in some cases abundance measures based on some regular grid-square system across the total area studied.

(2) Single species studies: the bird distribution for a single species is considered at the medium scale, e.g. birds on an island. Look-see methods may be used in which the observer searches for a low-density species based on *a priori* knowledge of its broad habitat requirements.

(3) Habitat-based studies: the bird distribution is considered at a small or minute scale by focusing on separate habitats, e.g. birds in a wood, on an estuary, on a heathland. Finest detail of scale can be obtained using radio-telemetry.

Atlas studies

The first major atlas of breeding birds was that undertaken by the BTO (Sharrock 1976). Sharrock developed the first system for achieving standardisation, and produced three breeding codes, possible, probable and confirmed, for each species in each 10 km square. An overview of grid-based atlas work is given by Udvardy (1981).

1. Considerations of scale

Atlas studies, for all species present, are conducted at a number of different scales. Obviously the smaller and more fine-grained the scale or grid, the more detail will be attributed to the bird data. Generally four categories of scale exist in bird distribution studies (minute, small e.g. at the scale of a wood, medium and large), and three of these are shown by examples in Box 9.2.

The scale of the study will be determined by the number of field workers available, the detail required, whether the objective is to estimate population size locally, regionally, nationally or internationally (or indeed at all). The capacity to relate to habitat data at the same scale may also be a consideration. The availability of maps is another important factor which is why some countries have used seemingly odd-shaped blocks. National atlasing of bird species distribution has become a major preoccupation of organised ornithology. A hierarchy of working groups, e.g. a national headquarters of an organisation coordinating on-the-ground field teams, is likely to achieve the best coverage and results.

Any atlas is organised on a grid basis. A grid of lines separated by 10 km, 2 km, 1 km, or other scale distance as appropriate and at right angles to each other, divides up the area to be atlased into 10 km, 2 km, or 1 km squares, etc. However, countries in which rectangular maps are produced generally adopt these for atlasing, rather than imposing the less familiar square grid system.

The following gives the scale of a number of previous bird atlases:

International

    Europe—UTM grid 50 km squares

National

    United Kingdom—10 km squares

    Netherlands—5 km squares

    France—Rectangles approximately 23 × 15 km

    Portugal—20 × 32 km rectangle

    Others in Europe—multiples of 5 or 10 km

    Madagascar—1/2 degree squares

    Tanzania—1/2 degree squares

    Uganda—1/2 degree squares

    Kenya—1/2 degree squares

    Lesotho—1/4 degree squares

    USA—State-wide e.g. 1 degree blocks, 5 km squares

    Canada—Province-wide, 10 × 10 km or 50 × 50 km

Regional

    Counties in UK—Tetrad (2 × 2 km) or 1 × 1 km in some.

2. Effect of grid size on species diversity

The number of species observed in a grid square increases with observation time and the size of the grid squares, which is directly related to scale. More species are discovered in large grid squares than in small ones, since the former are likely to contain more habitat types. Species diversity can be considered across a range of spatial scales, from continents to variations from point to point within a small copse or woodlot. Wiens (1981), using information from Whittaker (1977), describes seven diversity 'categories' in relation to an increase in area surveyed, or representing a change in diversity across an environmental or climatic gradient or between habitats (Box 9.3).

The relationships between species richness and grid size are non-linear, making it difficult to compare studies that relate to different-sized areas. Further, the number of species per unit area (D) and the proportion of grid squares occupied by a given species (i.e. grid-square frequency (F)) cannot be compared directly when grid sizes are different because the numerical figures change at a non-linear scale. Ellenberg (1985) presents a method of conversion in which plots of density and frequency of species in relation to grid size should be on a semi-logarithmic scale, so that there is a constant number of species added (or subtracted) for each duplication (or halving) of observation effort.

For a range of plot sizes between 10 and 1500 ha the number of bird species breeding in an area and the size of area measured in duplication steps i.e. 25 ha, 1 km, 4 km, 16 km squares, representing steps of 1, 2, 3, 4, respectively, are strongly correlated (Box 9.4). The constants of the linear regressions are characteristic for different broad habitat types.

Further, irrespective of the change of grid size (e.g. 25 ha to 1 km, or 1 km to 4 km), the relative change in F (grid-square frequency) is similar, enabling the calculation of conversion factors (CF) for F as a function of the grid-square frequency in a smaller grid square. Details are given in Ellenberg (1985).

3. Using historical information

Where resources do not permit a full and new survey, historical information is sometimes used. This has been the case for some African atlas studies, and in part for the European Atlas for countries with few birdwatchers. In such cases the historical data are in the form of bird reports for previous years, and of county gazeteers which may date, for Britain at least, back to the 18th century. These data have also been used to determine the changes in range of certain species in Britain, notably Buzzard, Capercaillie, Wryneck, Red-backed Shrike, Stone Curlew and Little Ringed Plover. Box 9.5 gives an example for the Red-backed Shrike. Usually, however, the assumptions and biases of the data collection are not known, nor is the sampling method, making comparisons of quantitative with descriptive data difficult to interpret. This approach can, however, be quite adequate for determining broad changes in range.

4. Planning an atlas

Many of the above atlases have endeavoured to use standardised methodology, e.g. the use of the same codes as the European Atlas for the Ontario Bird Atlas, and the recommendation that atlases in African countries should use the same scale as each other. The following are important considerations.

(1) The methods must be scientifically valid.
(2) The methods must be acceptable in the field to the largely amateur observers.
(3) The methods should be the same for all species, although some atlases do employ different methods for different species, particularly common versus rare ones.
(4) All data must refer to birds actually recorded. Observers must not be able to send in data based on what they 'know' to be present.
(5) The methods must be able to incorporate casual observations otherwise a great deal of potentially usable information will be lost, especially for the rarer and more elusive species.

Box
9.1

Patterns of distribution.

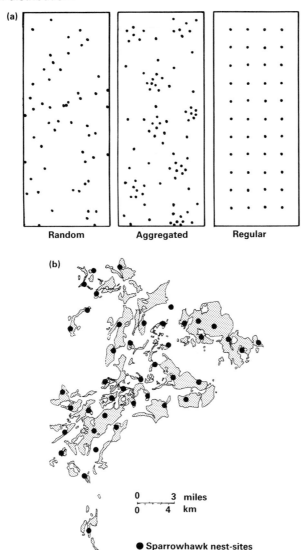

(a) Hypothetical examples of random, aggregated and regular distribution patterns (from Southwood 1978). Birds are rarely randomly distributed because the resources they exploit are rarely so dispersed. Clumped distributions are often observed, for example, colonial nesting seabirds (although even in this example, nests may be distributed regularly within the clump as a result of nest-defence e.g. Northern Gannet).

(b) Regularly distributed nest-sites. Distributions tending towards regularity are largely observed in species that defend a resource, for example, the distribution of nest-sites of the Sparrowhawk shows that they space their nests at a reasonably regular distance from each other at the landscape level (from Newton 1986).

Box 9.2

Different categories of spatial scale in bird distribution studies.

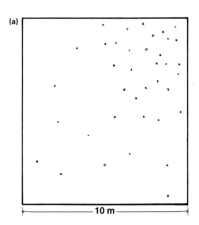

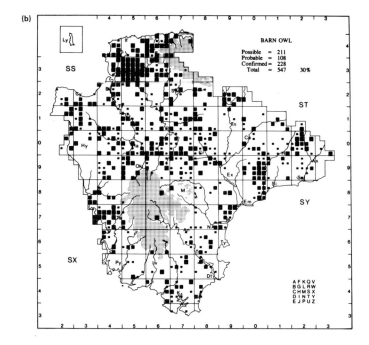

(a) Minute-scale distribution studies. This example shows the position of probe holes of waders in quadrats of 10 × 10 m placed on a muddy shore. A sample of these quadrat units across the shore would reveal patterns of differences in feeding intensity. For an example of a small-scale distribution study at the 5–100 ha scale, see Box 9.10.

(b) Medium-scale distribution studies. This example shows the distribution of the Barn Owl, based on tetrads (2 × 2 km squares) from the Devon County Atlas (from Sitters 1988).

**Box
9.2**
cont.

(c)

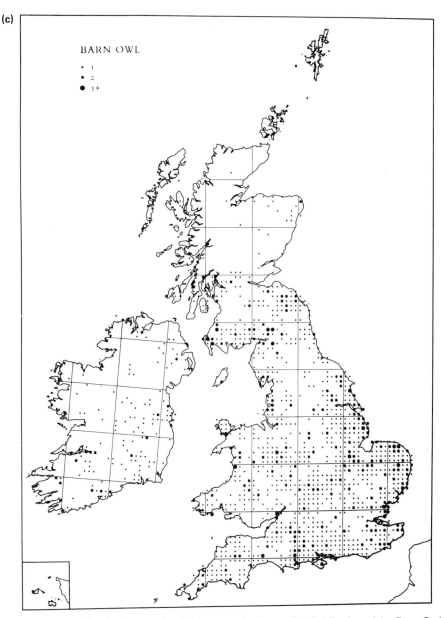

BARN OWL

• 1
• 2
• 3+

(c) Large-scale distribution studies. This example shows the distribution of the Barn Owl in winter, based on 10 × 10 km squares surveyed as part of the BTO Winter Atlas for the whole of Britain and Ireland (from Lack 1986). In Chapter 1 an example is given of what the Devon Atlas would look like had data been collected at the 10 × 10 km square scale, showing that much detail is lost with respect to the species relationship to land-use and habitat features.

Box 9.3   Levels and types of species diversity (from Wiens 1989 and Whittaker 1977).

Inventory diversities*	Differentiation diversities†
(1) For a small or microhabitat sample within a community regarded as homogeneous, subsample or point diversity	(2) As change between parts of within-community pattern, pattern diversity
(3) For a sample representing a homogeneous community, within-habitat or alpha diversity	(4) As change along an environmental-gradient or among different communities of a landscape, between-habitat or beta diversity
(5) For landscape or set of samples including more than one type of community, landscape or gamma diversity	(6) As change along climatic gradients or geographical areas, delta diversity
(7) For a broader geographical area including differing landscapes, regional diversity	

* Inventory diversity refers to that pertaining to a site at various scales of magnitude; essentially derived from a list and/or abundance measure.
† Differentiation diversity refers to that pertaining to a change associated with some gradient.

Effect of grid size.

**Box
9.4**

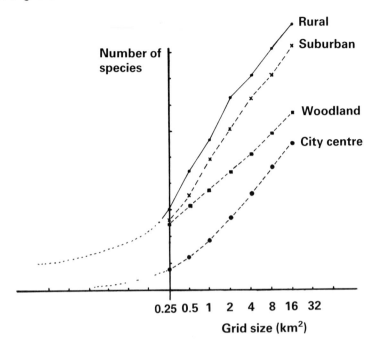

The number of species per unit area and the proportion of grid squares occupied by a given species i.e. grid-square frequency, cannot be compared directly when grid sizes are different because the numerical figures change at a non-linear scale. Ellenberg (1985) presents a method of conversion in which plots of density and frequency of species in relation to grid size should be on a semi-logarithmic scale, so that there is a constant number of species added (or subtracted) for each duplication (or halving) of observation effort. For a range of plot sizes between 10 and 1500 ha the number of bird species breeding in an area (on the *y*-axis) and the size of area measured in duplication steps (on the *x*-axis) i.e. 25 ha, 1 km, 4 km, 16 km squares, representing steps of 1, 2, 3, 4, respectively, are strongly correlated. As the plot size within the grid is increased in grid-based distribution studies, more species are encountered. The relationship is almost linear for a range of habitat types varying in structure. The main difference in the lines is due to the higher values of the intercept (on the vertical axis) for samples taken from more structured habitats. This shows that more structured habitats for a given area have more species, based on the assumption that greater structural diversity supports more exploitable niches.

Box 9.5

Long-term population monitoring using distribution studies.

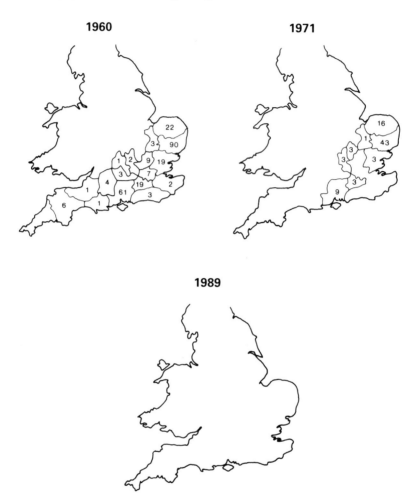

County recording of the distribution of pairs of Red-backed Shrike at irregular intervals since 1960 shows the decline in the species in southern England (from Bibby 1973). Here the number of pairs were added up by using county boundaries as units.

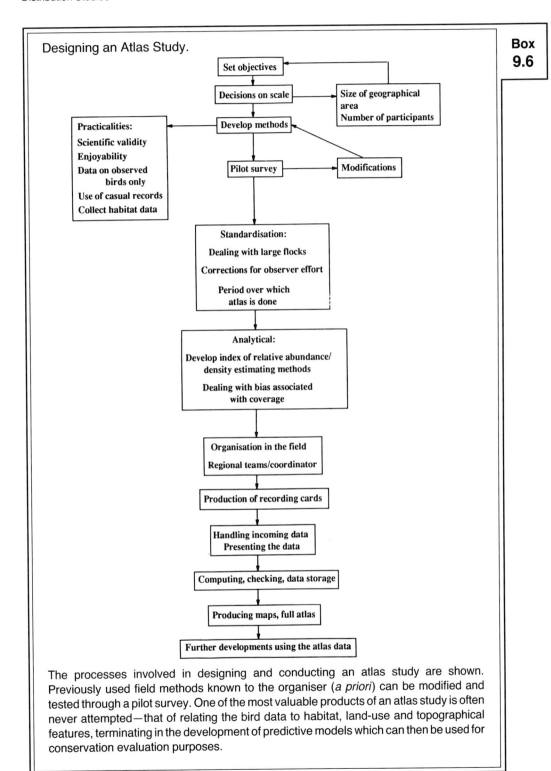

Designing an Atlas Study.

Box **9.6**

Set objectives

Decisions on scale → Size of geographical area / Number of participants

Develop methods

Practicalities:
Scientific validity
Enjoyability
Data on observed
    birds only
Use of casual records
Collect habitat data

Pilot survey → Modifications

Standardisation:
Dealing with large flocks
Corrections for observer effort
Period over which
atlas is done

Analytical:
Develop index of relative abundance/
density estimating methods
Dealing with bias associated
with coverage

Organisation in the field
Regional teams/coordinator

Production of recording cards

Handling incoming data
Presenting the data

Computing, checking, data storage

Producing maps, full atlas

Further developments using the atlas data

The processes involved in designing and conducting an atlas study are shown. Previously used field methods known to the organiser (*a priori*) can be modified and tested through a pilot survey. One of the most valuable products of an atlas study is often never attempted—that of relating the bird data to habitat, land-use and topographical features, terminating in the development of predictive models which can then be used for conservation evaluation purposes.

Box 9.7	Problems and solutions in designing the British and Irish Winter Atlas.	

Problem	Solution
Roosts	Counts made at roosts should be kept separate from other counts. After a pilot study the difference was not considered important
Habitats crossing 10 km-square	Deal with each 10-km square individually e.g. put a lake into a 10-km square in which it predominates
Determining position of square	In the case of estuaries square boundaries were boundaries described by natural points near to the real boundary
Birds flying over the square	Not included in the counts
Uneven coverage explains significant amount of variation in distribution e.g. lowland England received greater coverage than Scottish highlands	The maximum count of each species should be tested for correlation with the number of visit cards received for the square. If there is a correlation then uneven coverage could be biasing the bird distribution
Exaggerated impression of distribution of rare species	Caused by recording presence of one individual found in each of a widespread number of grid squares. Represent rare species independently from the main atlas maps by giving square-specific abundance measure
Flocks or rare individuals moving between squares	Difficult to overcome, apart from using ecological common sense at the regional or local level, e.g. Marsh Harrier recorded in three squares in the Wexford Slobs area in SE Ireland, all probably refer to the same individual

Sample positioning to relate distribution of a single species to large-scale habitat features. | **Box 9.8**

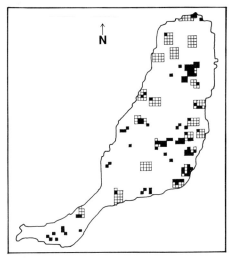

Using the 1 km grid map of Fuerteventura in the Canary islands, sample grid squares to be searched for the Fuerteventura Stonechat were determined randomly. Filled squares had breeding season records from 1985, 1984 or 1979; open squares were fully covered in 1985 but had no records of breeding chats. Topographical and habitat features of the squares with and without chats could also be recorded and used to relate, without bias, these features to bird abundance (from Bibby and Hill 1987).

Box 9.9

Distribution of a single bird species in relation to smaller-scale habitat features.

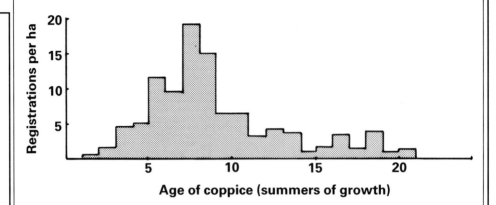

This figure shows per ha registrations of Nightingales in relation to the age of coppice (cyclically cut) woodland. The coppice age categories holding the highest densities of registrations are taken to be those preferred by Nightingales. This shows an example of a distribution study from which data are interpreted in relation to habitat type (from Fuller *et al.* 1989).

Aggregated distribution of a group of birds in relation to habitat features. **Box 9.10**

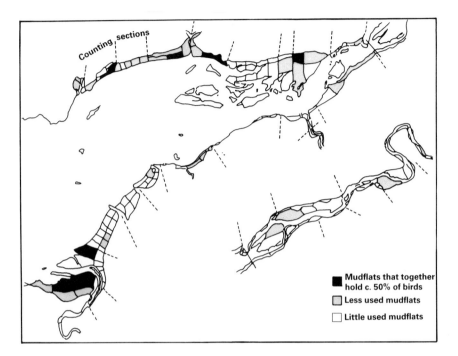

In this case the whole of the Severn estuary in England was divided into manageable units and wading birds (many species) were counted. The counts for all units were ranked and those making up a cumulative 50% of the total were shaded black. The map shows that the majority of birds are using only a small percentage of the total area of the estuary, which infers that resources also have an aggregated distribution. Note that, compared with Box 9.2, this example represents a small-scale distribution study (from Clark 1990).

**Box
9.11**

Duck nests in relation to vegetation types.

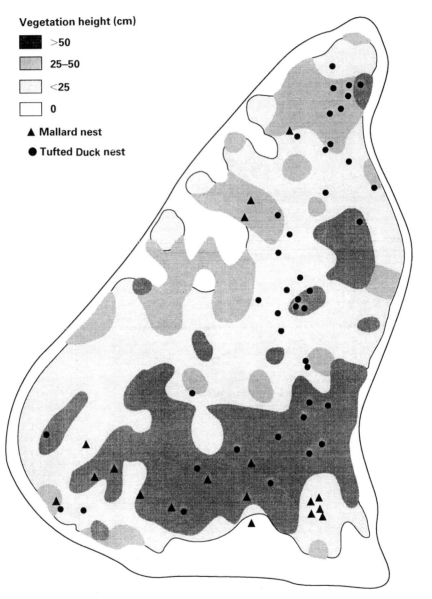

Vegetation height (cm)
- >50
- 25–50
- <25
- 0

▲ Mallard nest
● Tufted Duck nest

In this example vegetation height was measured at each intersection of a grid of 5 × 5 m placed across the island on Willen Lake in Buckinghamshire. Four height categories are represented on the map by different shading, having first interpolated similar heights into contours. A plot of Mallard and Tufted Duck nests is overlaid on the vegetation map. By calculating the area of each vegetation height and relating to the observed and expected (based on area) number of nests, selection for the tallest height category is demonstrated (from Hill 1982, Hill 1984b).

**Box
9.11**

cont.

The number of Mallard and Tufted Duck nests is shown in patches of vegetation of different heights. (Chi-square comparing observed with expected nests $= 18.8$, degrees of freedom $= 3$, $n = 62$ nests, $P < 0.005$). Nests were placed in taller vegetation than would have been expected on the basis of the area of habitat available (from Hill 1982, Hill 1984b). The two species were analysed together since their fate (e.g. predation) operates on the population of nests rather than the nests of the two species independently, since their breeding seasons overlap.

Vegetation height (cm)	Area of vegetation of this height (ha)	Observed nests	Nest density (no. per ha)	Expected nests
>50	0.83	26	31.3	15.3
25–50	1.33	23	17.3	24.5
<25	0.52	12	23.1	9.6
0	0.68	1	1.5	12.5

Box 9.12

Radio-telemetry used to determine distribution of individuals.

(b)
Radio-locations

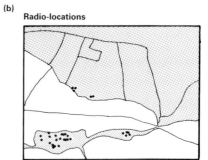

Minimum polygon area

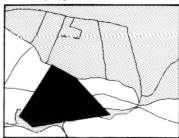

85% multinuclear clustering

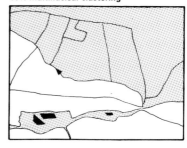

(a)

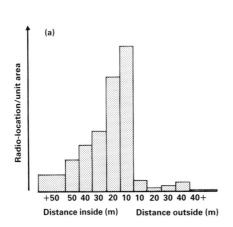

Radio-location/unit area

+50   50 40 30 20 10 10 20 30 40 40+
Distance inside (m)   **Distance outside (m)**

(a) Radio-locations to determine habitat preference. Radio-locations of Pheasants, assigned to the distance of the bird from a woodland edge (both inside and outside the wood), show a significant preference for edge habitats. One immediate interpretation would be that the Pheasants, feeding just outside the edge of the wood, move into the edge on the approach of the radio-tracking observer. However, locations were taken at an adequate distance to overcome this problem (from Hill and Robertson 1988).

(b) Radio-locations used to determine home range. In the first example, the minimum polygon area (MPA) method is used to determine the home range of Pheasants; in this method the outermost points are joined together. This gives a totally different value to that when home range is calculated by the multinuclear clustering method, producing a probability contour. The MPA method fills in more non-used habitat than the others (from Robertson *et al.* 1990). See also Kenward (1987) and Dixon and Chapman (1980) for the harmonic mean contour method.

**Box
9.13**

Distribution data used to calculate population levels.

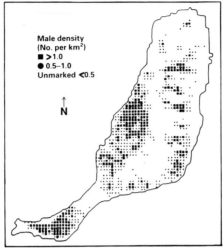

Male density
(No. per km²)
■ >1.0
● 0.5–1.0
Unmarked <0.5

↑
N

The topography of Fuerteventura was measured by the number of 20 m contour intersections of the boundaries of each 1 km square, and related to the number of male chats per square km. The regression line fitted to the original data is $y = 0.027 + 0.024x$, $r_{194} = 0.342$, $P < 0.001$. This regression was then used to predict the distribution map of Fuerteventura Stonechats on the map-derived measurement of topography of the whole island (from Bibby and Hill 1987).

**Box
9.14**

Distribution data used to estimate the British population of Wood Warblers (from Bibby 1989).

Region	Scotland	Lakes	Marches	Wales	South west	Rest of England	Total
10-km squares covered	69	11	21	47	26	243	
Total birds recorded	749	215	290	1636	610	1936	
Mean birds per square	10.9	19.6	13.8	34.8	23.5	8.0	
Total squares in region	317	43	71	159	83	430	
Estimated population	3441	840	980	5535	1947	3426	16 169
Upper confidence limit	4020	1130	1280	6390	2390	4160	17 560
Lower confidence limit	2880	590	730	4720	1550	2860	14 850

(6) Two or more observers will sometimes be working independently in the same area. The methods must not involve a subjective decision as to which observations to use.

Principles broadly relate to assessing objectives, deciding on scale, developing methods, standardisation, analysis, organisation in the field, computing and producing maps. A diagrammatic representation of the process of designing and conducting an atlas study is given in Box 9.6.

Examples of atlases

1. The atlas of wintering birds in Britain and Ireland

This example is given to show the various stages in the process of producing an atlas of bird distribution. A full-scale pilot survey for the Winter Atlas (described in more detail in Lack 1986) was conducted in the winter 1980/81 with two main aims. Firstly, it was necessary to find a method of assessing abundance; secondly, it was necessary to find out about any movements between November and March in order to define the limits of the 'winter', i.e. the months that can be categorised most specifically as winter based on their bird complement. It was decided not to start the field season until the middle of November, and to finish at the end of February before breeding activity commenced in order to reduce effects caused by movements of migrants.

The pilot survey suggested that the number of birds seen on any one day was a good unit of relative abundance. The 'day' was standardised as a period of 6 hours (see below). It was also decided to take the maximum number of birds counted on one day as the measure of abundance. Sometimes this might lead to one particularly large count being used, but this risk was outweighed by major statistical difficulties when calculating means or medians, caused particularly by casual records.

This method has two weaknesses.

(1) It is possible to use misleadingly one particularly large count, e.g. a flock.

(2) There is only correction for observer effort within a day but not for the total number of days, which may have specific species as well as overall community biases.

The strengths of this method are associated with difficulties when calculating means or medians, particularly so in the case of casual records. For example, a 'zero' count for a species in a restricted habitat, e.g. ducks on a lake, could mean either that there are no ducks there or that the habitat was not visited and yet ducks were present.

Two kinds of records were accepted: first, the result of a visit to a 10 km square specifically to do field-work for the Winter Atlas; and second, any casual records (termed Supplementary Records) of individual species.

For a specific visit observers were asked to spend a minimum of 1 hour in their 10 km square and to count all birds seen and/or heard. At the end of the visit the total number of each species was recorded on a Visit Card to aid easier computer checking, together with the 10 km square, an identifying feature of the square, the date, the time spent in the field and the total number of species recorded.

A 'day' was defined as 6 hours in the field, this being the longest that most people would be likely to spend doing field-work on a winter's day. In fact only about 3.5% of all cards received were for periods longer than this.

All timed counts of longer or shorter than 6 hours were standardised to 6 hours to permit better comparisons of areas that might have only 1- or 2-hour counts with those that had 6-hour counts. The procedure adopted was to calculate, for each species individually, a coefficient for the regression of number of birds seen on time spent in the field. The data were normalised by putting both axes on a logarithmic scale. With a large number of data points available even quite weak relationships between numbers of birds and time spent in the field are significant at the usual statistical point of $P = 0.05$. As nearly 200 species are considered in the atlas, standardisation corrections were used only if the relation was statistically significant at $P < 0.001$. Many of the commoner land birds came into this category. The majority of the rarer species and those restricted in their habitat preferences have a zero coefficient and no corrections were made. In practice this means you are just as likely to see a rarer or more elusive bird in the first hour of a count as in the sixth or, similarly, visit the restricted habitat, e.g. a lake, in the first or sixth. You are unlikely to accumulate more and more as field-work continues. For the commoner land birds though this is what does happen, and therefore the coefficient is positive and standardising corrections are needed by multiplying by $(6/T)^b$ where 6 is the standard 6 hours, T the actual time spent on the count and b the regression coefficient.

The final atlas (as with the first Breeding Atlas by Sharrock) was accompanied by overlay maps of topographical and environmental information, to aid the reader's interpretation of bird distributions.

Problems encountered with the Winter Atlas and their solutions are given in Box 9.7.

2. The atlas of wintering North American birds

This atlas is an analysis of the Christmas Bird Counts conducted since 1900 (Root 1988). Assumptions and refinements to the counts are given in Bock and Root (1981), Drennan (1981) and Arbib (1981). Each count site covers a circle of 15 mile (24 km) radius and at least 8 hours must be spent counting. Twelve hundred or so sites are covered annually on any day within a 2-week period around Christmas. For the purpose of the atlas, mean values counted per site per year over a 10-year period were used to produce computer-generated contour and 3-dimensional maps of distribution and abundance patterns of species in winter. The use of means over a 10-year period is thought to reduce any spurious effects due to weather and abnormal movements of birds. These means summarise the raw data and are therefore an interpretation.

There are a number of deficiencies in the counting procedure. Sites at which counts take place are not uniformly distributed so that there may be biases due to uneven coverage. The abilities of participants, the miles they travel, the hours they spend counting and the size of counting parties, differ between count sites. Further, abundances of gregarious species are inaccurately recorded for two reasons: because flocks are difficult to count accurately; and because chance movements of large flocks can significantly change the recorded abundance of a species.

Variation in count effort at the different sites was diminished by dividing the number of individuals seen at a site by the total number of hours spent counting by the groups of people in separate parties at a site. Mean values at each site were then calculated by summing these values over the various years and dividing by the total number of years the count was held. Since the area counted is restricted to a 15 mile radius, these mean values are densities.

The density values are normalised to range between zero and one for each species by dividing the mean values at each site for a given species by the mean value at the site with maximum abundance. These normalised values are plotted. Maps of species with extremely high (more than 200 individuals counted per hour) or low (fewer than 0.2 individuals counted per hour) abundances are excluded from the main section of the atlas because of difficulties over their interpolation. The atlas also presents overlay maps of elevation, vegetation, mean minimum January temperature, mean winter ocean surface temperature, mean length of frost-free period, mean annual precipitation, general humidity, and national wildlife refuges. Although there are a number of biases and difficulties with the North American Atlas, the vast area covered and the participation required precludes a more statistically valid project being undertaken cost-effectively. This atlas is a good example of the potential of amateur field-work.

3. The atlas of the birds of the Netherlands

This atlas was constructed by the Netherlands' ornithological body SOVON (SOVON 1987), from data collected monthly from October 1978 to September 1983 in 5 × 5 km grid squares. Birds both using and flying over the squares were recorded simultaneously, the former often being accompanied by numbers observed. The monthly distribution maps represent a cumulation of 5 years field-work, so that, for example, a map for January contains the results for the Januarys of 1979, 1980, 1981, 1982 and 1983. For most species histograms are used to illustrate their occurrence throughout the 60-month period of field-work. The columns of the histograms represent the proportion of squares in which the species was observed in the month concerned, corrected for the number of squares observed per month. Because it has been conducted at the 5 km square level over a 60-month period, this atlas represents one of the most intensively detailed yet attempted in any country. Overlays of ecological data are also provided at the same scale as that for birds, detailing deciduous forest, coniferous forest, coastal dunes and beach, wet moorland, heathland, marsh, standing water bodies and drift sands.

4. The raptor grid scheme of Finland

The scheme, begun in 1982, aims to collect data on population size and
nesting success of Finnish raptors and owls, to establish population trends,
and to use nest-site data for conservation purposes. A minimum of 200 hours
per raptor grid square (10 × 10 km, chosen freely by a birdwatching group)
is spent on field-work during February to August. In each square aerial
displays of diurnal birds of prey are noted (April), owls are listened for
(March), and nests and fledged young are searched for (owls—May and
June, respectively, hawks—June and July, respectively). The data are stored
in map form thereby constituting a distribution study of both abundance and
nesting success, which can then be interpreted geographically.

Single-species studies

Distribution studies of single species may require a more fine-scale approach
than the larger-scale atlas studies of all species described above, but this
depends on the original objectives of the study. It is important that the
reasons for conducting a single-species survey are identified at the outset and
that the methodology maximises the potential for interpretation, for example
the calculation of densities and their relation to habitat variables.

The first example given is that of the Fuerteventura Stonechat, one of the
most localised bird species occurring in the Western Palearctic and endemic
to the arid island of Fuerteventura in the eastern Canaries. This island was
visited for 16 man-weeks from 18 February to 11 March 1985 with the aims
of estimating the numbers and distribution of the bird, describing its habitats
and assessing its likely future welfare (Bibby and Hill 1987).

Fuerteventura, with an area of 1653 km^2, is sparsely vegetated and moun-
tainous but also has stony plains and sand dunes. The survey was based on
21 blocks each of 12 1 × 1 km grid squares (Box 9.8). Inland the blocks were
3 × 4 km rectangles. The central squares of the blocks were selected randomly from a list of all squares on the island. If surrounding squares
contained no land, an adjoining terrestrial square was selected. The sampling pattern was determined in advance and bore no known systematic
relationship to physical features, vegetation or likely suitability for chats.
Blocks were visited only once by field teams working at an intensity of about
2–3 man-hours per km^2. Pairs had nests, or, in many cases, fledged young
and were generally noisy and conspicuous in open terrain.

For the purpose of analysis, 1 km squares were excluded if they had not
been completely covered for lack of time. Some of the random blocks were
less than 12 km^2 in area because they adjoined the sea. The 21 blocks had a
land area of 235.2 km^2 of which 209.8 km^2 (89%) was searched. The results
are a random sample of 12.7% of the land area of Fuerteventura. A later
section of this chapter describes how the information was used to estimate
population size.

A special case of single-species studies applies to rare or low-density
species using look-see methods. The amount of effort required to sample the

species using random sampling would be inappropriate and too costly of time in such a case. Using *a priori* knowledge of the species requirements, the method entails searching the most likely areas in order to maximise the rate of encounter of the species. Habitat measures taken at localities where the species is present can be paired with sites in which the species is absent.

Habitat-scale studies

Mapping the distribution of species in specific habitats is done for two main reasons: (1) to relate the distribution of bird registrations derived from territory mapping to the availability of habitat types, thereby enabling preference or avoidance to be determined; and (2) for specific management objectives.

1. Bird registrations and habitat types

The distribution of Nightingales (and other songbirds) in Ham Street Woods, Kent, an area of actively coppiced woodland, shows a preference for 6–7 year coppice stands than for older and younger types (Fuller *et al.* 1989). In this study, birds were counted by a single observer using territory mapping (see Chapter 3), over a 5-year period. Each year 23–25 visits were made, spread throughout the breeding season from late March to early July. Each year the entire site was covered evenly and coverage (effort) was consistent between years. One important modification to the methodology was that, because the site, with 52 compartments (blocks of woodland), was extremely complex, it was impossible to assign most of the 'territories' to one compartment or another. Therefore, the densities of 'registrations' recorded in compartments of different ages were used as indices of abundance. The map of the registrations for Nightingales is shown in Chapter 10.

The age of coppice (number of summer's growth) for all compartments was known and used to investigate the density of registrations in coppice of varying age (Box 9.9).

For each year class of coppice the total number of registrations recorded in that year class was divided by the total area. The resulting index was expressed as the number of registrations per ha. There are two potential problems with interpreting the derived patterns of bird distribution: (1) individual compartment effects (e.g. coppice composition, soil type) could be confounded with year-class effects, but as compartments within each age class were widely distributed this was not too severe; (2) it is not possible to test statistical deviation of the patterns away from randomness because the registrations upon which they are based are not independent samples but are, in many cases, repeated observations of the same individual. This could be overcome if results were based on territories.

Roost and feeding sites of waders on estuaries provide another example of distribution within a habitat. There may be fine-grained environmental changes across the estuary which drive the distributions of birds. Such plots reveal the aggregated distribution of waders on some estuaries in which honey-pot areas hold most of the birds. Box 9.10 shows an example for the Severn estuary.

2. Nest distribution patterns

We have already given an example of the use of maps of nest-site locations in explaining the distribution of breeding Sparrowhawks (Box 9.1b). Here a further example is given of the use of nest distribution patterns in relation to vegetation for Mallard and Tufted Duck. Regular searches of a small island on a lake in Buckinghamshire were conducted to find all duck nests and to monitor their fate. Vegetation height across the island was also measured. The resultant contours of vegetation are shown in Box 9.11 overlaid with a map of the nest positions. Observed nest density is compared with that expected on the basis of no selection for height. A chi-squared test was performed (Box 9.11) but a more rigorous test would have been that of Neu *et al.* (1974).

3. Radio-telemetry

Distribution data obtained by radio-telemetry are less biased by the observer than those collected as part of a survey count or census, and are particularly useful for determining micro-habitat selection by a species, such as selection of a small group of shrubs or of one crop type in preference to another. Firstly, the radio-locations are marked on a map (or input to a computer database which can generate them in relation to a study area map), and the home range is then determined by joining the outermost points (minimum polygon method) or by some probabilistic method, such as by calculation of the harmonic mean or multicentred clustering, which identifies core activity areas (see Dixon and Chapman 1980; Kenward 1987). Box 9.12 gives an example of the use of radio-telemetry in studying habitat selection by Pheasants (Hill and Robertson 1988). An example of a multicentred cluster-ing and of a minimum polygon area plot for the same data from the Pheasant study are also shown, to illustrate the different areas produced by the two methods (Box 9.12).

  The amount of use of a habitat as determined by radio-telemetry, or the numbers of registrations of birds within it, can be related to the availability of that habitat using various preference/avoidance indices (Ivlev 1961; Jacobs 1974). Such indices provide only a ratio of habitat use to habitat availability and do not provide statistical test. A number of other models use a common statistical approach; Alldredge and Ratti (1986) compare four such tech-niques (Quade 1979; Neu *et al.* 1974; Iman and Davenport 1980; Johnson 1980). Most use a chi-squared goodness-of-fit analysis to test whether obser-vations of habitat use follow the expected pattern of occurrence based on habitat availability.

  The models test slightly different hypotheses and a number of the follow-ing assumptions are applicable to different models. (1) All observations are independent in that location 'fixes' are collected far enough apart in time for there to be no temporal serial correlation, i.e. the presence of a bird at position x at time a should not affect its position at time b. Technically, over short periods of time the observations of the same bird cannot be independent. Furthermore, at its worst, numerous radio-locations are really pseudo-replicates. (2) The sample size is sufficiently large to allow a chi-square approximation for the goodness-of-fit statistic (i.e. more than 1 expected observation in each habitat category and less than 20% of all categories contain fewer than 5 expected observations. (3) Habitat availability is the same for all individuals. (4) Results from one animal do not influence results from the other animals. When the number of habitats is small, and with more than 20 animals and 50 observed locations per animal, the Neu method performs best. One warning—with all methods, as the number of habitats used increases, the multiple comparison error rates increase; therefore the number of habitats considered should be limited in the study design.

For radio-tracking data, in which visually obscured individuals can be tracked without disturbance and hence bias, the area of each habitat in the core areas (or 85–95% range contours) can be calculated and the observed number of locations in the area compared to that expected if the bird wandered randomly throughout the area defined by its home range contour, polygon, or some pre-determined sample area. The use of compositional analysis (Aitchison 1986; Aebischer & Robertson 1992), overcomes various constraints posed when habitat use is presented as proportions and analysed by chi-square methods. Compositional analysis involves using a 'logratio' transformation on the proportions of different habitat types available and used by the bird. One of the principles underlying the method is that all habitat types should be considered simultaneously because the results from analysing one type at a time are not independent, simply because if one habitat type is used a lot, at least one other will be used less as a consequence.

Examples of the use of distribution studies

1. Estimating population size

Three examples are presented in which population size is estimated from well designed distribution surveys.

The first is for the Fuerteventura Stonechat, for which the study design has already been described. Total population estimates were made from the number of known pairs and from the total numbers of males (pairs plus extra males). Densities varied considerably between plots, and since each was a different area, variances of the estimated total populations were calculated by jackknife (Miller 1974) thus:

Proven pairs: mean = 591, 95% confidence range = 500–682
All males: mean = 779, 95% confidence range = 663–893

If this survey was 90% efficient and half the extra males found actually represented breeding pairs, a round estimate for the population in 1985 would be 750 ± 100 pairs.

Within the formally randomised census, 196 non-coastal squares were searched and the numbers of males per square were regressed on land-form factors measured from maps. In particular there was a strong correlation between number of males per square km and topography, as measured by the number of 20 m contour intersections of 1 km square boundaries. This model was then used to estimate a total population of 880 males (95% confidence range 792–962) for the whole island based on the whole island's topographical values, and using the model, the distribution map for the chat was predicted (Box 9.13).

The second example is for the Wood Warbler breeding in Britain (Bibby 1989). Wood Warblers have a very distinctive and carrying song and are readily detected so there is close coincidence between single visit survey results and actual numbers (Bibby 1989). These birds therefore lend themselves well to a single-species survey based on fewer visits than is involved in mapping studies. The objective of the survey was to count a random sample of one-third of those 10 km squares with proven or probable breeding in the Atlas of Breeding Birds (Sharrock 1976) in northern and western regions where full cover was not thought achievable. For the rest of England it was intended to count all such atlas squares.

Observers were asked to make a single visit to all suitable habitat in each chosen 10 km square during 15th May to 10th June. Counts were returned as totals by tetrad, or else tetrads were recorded as unsuitable for these birds. The total number of singing birds in those 10 km squares in which the species probably or certainly bred in the period of the atlas was estimated by assuming that the samples covered were a random selection.

Because of the non-normal distribution of counts and differing sample fractions between the different regions of Britain, Monte Carlo simulation (Buckland 1984) was used to calculate confidence limits, by taking the average for each statistic from 10 simulation runs, each of 500 trials. The estimate for the whole of Britain using these techniques was 16 170 singing birds, with fairly wide confidence limits (14 850–17 560) (Box 9.14). The range is high because the number of birds in a square varied widely (from 0 to 253) so it is difficult to be confident about estimating numbers in squares not counted. Further refinement, giving an estimate of $17\,200 \pm 1370$, was achieved by including 'possible' atlas records.

The final example is for the breeding Lapwing survey conducted in Britain in 1987 (Shrubb and Lack 1991). First, every 10 km square in England and Wales that contained some land was identified. Within each a tetrad was randomly chosen, even if the tetrad within the 10 km square fell on water. Observers were asked to visit these chosen tetrads and count the numbers of breeding pairs (males displaying, females incubating, pairs occurring together) of Lapwings by looking into every field in that tetrad. Habitat type was also recorded. One figure for the total number of pairs in the tetrad was produced. The counts were found to be skewed, with 60% of counts having zero Lapwings, if one includes 'counts', for tetrads located in the sea. The sample counts were multiplied by 25 to give the number of pairs in each 10 km square. This was done regionally (nine MAFF regions), and the regional values were summed to give a figure for England and Wales. Because the sample counts were skewed, 95% confidence limits were calculated by bootstrapping methods (Efron 1982).

2. Relating distribution to environmental data

This is one of the most important uses of distribution studies. Grid-based environmental data are available from a number of sources, and can be related to bird data at the appropriate scale, for example as that collected as part of an atlas study. Information from the following UK-based environmental databases could be related to bird data using multivariate statistics.

(1) Land Characteristics Data Bank collected by the Institute of Terrestrial Ecology in the UK.

(2) Land Classification System also collected by the Institute of Terrestrial Ecology.

(3) Agricultural Census Statistics collected by the Ministry of Agriculture Fisheries and Food in England and Wales and the Department of Agriculture and Fisheries in Scotland.

Together these databases provide measures of habitat, topography, climate and land-use at various scales down to 1 km^2. Predictive models could be constructed which incorporate these data with bird data in order to identify possible effects of land-use change on species (and communities).

An example of the use of such statistics is that for the BTO 1987 Lapwing survey. Details of the survey are given above under the section on estimating population size. For the purpose of identifying habitat selection by Lapwings it was assumed that the proportion of various agricultural habitat types within the sample tetrad was the same as that for the 10 km square. Habitat proportions in each region were obtained from MAFF statistics, and these were used to calculate the expected number of Lapwings for each of the nine MAFF regions. The expected numbers of Lapwings in each habitat type were then compared to those observed using a preference index. It is also possible to carry out the same analysis within individual regions as opposed to nationally in order to identify strong regional preferences in crop types.

The scale at which the bird data are collected will influence the predictive value of models so developed. For the Devon Atlas (Sitters 1988) for example, tetrad-based data on geology, extent of urbanisation, agricultural land quality, woodland coverage, and location of standing water would enable a useful multivariate analysis of bird distribution in relation to these variables. This approach could be used to validate models built from other data collected at a larger scale.

Osborne and Tigar (1992) used logistic models of bird atlas data (presence/absence) in relation to topographical and land-use data for Lesotho, South Africa. The probability of finding a species in an unvisited grid square enabled prediction of the distribution of species, thereby allowing gaps to be filled in on the basis of the model predictions.

3. Conservation evaluation

Species distributions can identify areas or tracts of land that might benefit from site protection and conservation through one of the various statutory legislative forms of designation. This is relevant in the UK, for example, in the identification of 'Special Protection Areas', 'Environmentally Sensitive Areas' and 'Sites of Special Scientific Interest' etc., administered by the British government's various conservation and land use agencies. In terms of numbers of species the Breeding and Wintering Atlases would enable regions to be identified with respect to their species-richness. Map-based information is presented using symbols increasing in size to indicate greater species-richness. In winter, coastal regions, particularly those in the south of Britain, are more species-rich than sites further inland, whereas the uplands are seen to be species-poor. Splitting the data into species of similar groups such as freshwater species, waders and seed-eating species reveals, however, major differences in distribution. *A priori* knowledge of a species' or group's ecology can allow biologically meaningful splitting of the data in this way.

The Winter Shorebird Count organised by the BTO is an example of a distribution study with delimited boundaries which has special reference to conservation evaluation. The whole coastline of Britain outwith estuaries was walked by a team of observers at low tide between mid-December and mid-January. The areas were divided into segments on the basis of their primary habitat type—bedrock, boulders, cobbles, gravel, sand and mud. Secondary habitat types, including all of these plus weed cover, and slope of the shore, were also documented. The bird counts can thus be analysed with respect to habitat preference, and evaluation of particular stretches of coast-line based on species and community assemblages can be made.

4. Effects of weather

Distribution studies, such as those of the Breeding and Winter Atlases have been used to investigate the effects of weather on such subjects as (1) early breeding in a number of species, (2) mortality in cold winters, (3) body size and winter distributions, (4) seasonal patterns of movement.

An increase in the number of 10 km squares in which the species was recorded in late winter during the Winter Atlas showed evidence of early breeding activity in Corn Bunting, Lesser Spotted Woodpecker, Goshawk, Hawfinch, Dipper, Raven and Golden Eagle. Because the extra registrations were randomly distributed throughout the ranges, the increases were probably not due to movement of the population. However, caution is needed in such interpretations. In other species marked declines for the same periods were noted, indicating higher mortality in cold weather and towards the end of the winter. Kingfisher, Grey Heron and Goldcrest are well known to be susceptible to cold weather.

The Winter Atlas was also used to investigate how species of different body weight are distributed during winter. Bigger species tended to be distributed further north than smaller species.

Seasonal patterns of movement involving the arrival and departure times of winter migrants and the arrival and departure times of breeding migrants into Britain have also been studied using the Winter Atlas. Breeding-ground conditions for northern breeding waders such as Dunlin, and geese such as Brent Goose, influence the time of arrival in Britain where they will expect to spend the winter. This has been shown using comparisons of distribution maps for winters varying in their severity. Likewise, warm conditions towards the end of winter hasten the departure of these and other winter visitors and encourage the early arrival of spring migrants such as warblers and hirundines. Upland breeding individuals of species that breed across Britain such as Song Thrushes tend to move to the coast in winter. This was determined by analysing seasonal patterns of distribution from the Winter Atlas (Lack 1986).

5. Identifying partial migration

The Winter Atlas illustrated the two types of partial migrants in Britain and Ireland. (1) Some individuals leave Britain and Ireland in the autumn while other individuals remain. (2) The breeding population is augmented in the winter by birds from Fennoscandia and other parts of northern and eastern Europe. Males of certain species, e.g. Chaffinches and Pochard, are more common in the north than in the south of Britain, so there are partial migrants with respect to sex. A number of species, for example Stock Dove and Skylark, in the Winter Atlas were shown to vacate the north of Britain before the end of the winter period. These movements are thought to be in response to food availability. This information is a valuable use of atlas data in identifying partial migration through temporary changes in distribution.

6. Irruptive species

The Winter Atlas also showed patterns of distribution of species that irrupt through, generally, shortages in their food supplies in Scandinavia and elsewhere in northern Europe. For Britain there are four classic irruptive species—Waxwing, Crossbill, Brambling and Siskin. From the Winter Atlas the species were found to irrupt in different years probably because they feed on different foods. The analysis of irruptions is based on counting the number of squares in which the species is recorded in different years.

Summary and points to consider

What questions are to be addressed by the distribution study? Set objectives.
What is the appropriate scale?
Is the study appropriately atlas, single species or habitat-based?
Are habitat data to be collected alongside the main survey data?
Is a pilot study necessary to identify problems?
Make sure the methodology is standardised (same grid size as a previous study in order to allow comparisons).

Make sure the methodology is adequate to achieve the objectives e.g. to be able to place confidence limits on population estimates.

Has the effect of grid size on the number of species observed been taken into account in the study design?

How will coverage problems be dealt with?

How are data to be handled and dealt with? Design the data collection in a way that eases the methods of analysis.

What statistical treatments are necessary?

10

Description and Measurement of Bird Habitat

Introduction

Counts of birds in a study plot become more valuable if they are related to the habitat variables within the plot, as such analyses can discover factors which affect bird occurrence or abundance (see MacArthur and MacArthur 1961; Cody 1985; Rotenberry 1985; Wiens 1989). Moreover, understanding bird–habitat relationships can help predict the effects of management and habitat change on bird populations.

The level of detail collected on the habitat(s) to be studied should be related to the objectives of the study. If, for instance, the distribution of birds over an extensive area is being assessed, habitat information at the broad level (perhaps from satellite imagery or aerial photographs) may be all that is required. However, investigations aiming to elucidate habitat preferences of a particular bird species require more detailed and time-consuming studies of the birds' habitat, often involving the measurement of habitat variables in sample plots, or at the exact position of the bird. In such cases the collection of data on the abundance of a particular bird and variation in the habitat in which it lives is often followed by data-analysis using multivariate statistical procedures (see Gauch 1982).

In general, there is a gradation of detail of habitat recording in ornithological studies from the broadest scale where habitat details are mapped and the positions of birds are marked on the map (map based), to a finer scale where the bird populations and habitat variables are sampled in a statistically representative number of sites (plot based), and the finest scale where habitat variables are recorded at the exact position of a territorial or radio-located bird (individual based) (Box 10.1).

In this chapter, this gradation of scale is described and illustrated with examples taken from Great Britain and North America.

Habitat mapping methods

Mapped counts of birds are fairly meaningless without an adequate mapping of habitat features in the area (see Chapter 1, Box 1.2). Hence, a typical first step in understanding the habitat preferences of birds in a study area is to produce a map showing the habitat features.

Habitat maps are generally produced in several stages.

(1) A base map of the area is obtained. Such maps may be national cartographic maps produced at various scales, aerial photographs taken from light aircraft or balloons, or photographs produced from satellites such as the Landsat system.

(2) A provisional habitat map of the study plot is traced from the base map onto tracing or similar paper using fine-nib pens. In Britain, provisional habitat maps are commonly drawn at 1 : 10 000 or 1 : 2500 scale depending on the size of the study area and the spatial resolution needed for plotting data. The boundaries of the study plot and obvious major divisions or features such as roads, woodlands, built-up areas or arable farmland, as well as reference points such as isolated farm houses, are marked.

(3) The provisional habitat map is checked and refined by a ground survey of the study area. The level of detail mapped in at this stage should be related to the problems being addressed by the study. It is important that the maps are not too detailed as the time taken in their production will be wasted, or too generalised as little will be learnt from their use.

In Britain, a broad scale of habitat mapping can create habitat maps according to standardised habitat types (e.g. Fuller 1982). The following 26 bird habitats in Britain are defined by Housden *et al.* (1991).

Natural and semi-natural
    (1) Montane
    (2) Upland heaths
    (3) Upland mires
    (4) Upland grasslands
    (5) Broad-leaved woodland/scrub
    (6) Native pine woods
    (7) Lowland heaths
    (8) Downland
    (9) Swamps/fen/carr
  (10) Lowland wet grasslands
  (11) Marine
  (12) Inshore waters
  (13) Sea cliffs and rocks
  (14) Intertidal flats
  (15) Salt marshes
  (16) Shingle/sand/machair
  (17) Coastal lagoons
  (18) Oligotrophic/mesotrophic waters
  (19) Eutrophic waters
  (20) Rivers and streams
  (21) Marine
Mainly artificial
  (22) Plantations

(23) Extraction pits and reservoirs
(24) Arable
(25) Improved pastures and leys
(26) Built-up areas

These habitats all have distinctive bird communities. An advantage of this system is that habitat divisions can be easily and rapidly recognised by non-specialists, hence the classification is cheap in labour terms. Disadvantages are that only the most broad bird/habitat relationships can be described (Box 10.2).

In Britain it is also possible to define the habitats present on a study plot to a much finer degree using hierarchical habitat-classification systems. Habitats can be described in terms of the Royal Society for Nature Conservation/ Joint Nature Conservation Committee (RSNC/JNCC) system, which aims to give an alphanumeric code to all British habitats (NCC 1990). Vegetation communities can also be defined throughout Britain using the system of 'British Plant Communities' (e.g. Rodwell 1991).

Box 10.3 presents an example of the hierarchy for woodland and scrub habitats. An example hypothetical woodland could be coded as A.1.1.1.(W8), which would translate to: a broad-leaved, semi-natural, high forest of the *Acer campestre–Fraxinus excelsior* woodland vegetation community. Given sufficient data, habitats can be described down to the vegetation community level throughout the rural and urban environment of Britain.

Non-quantitative field survey is adequate to ascribe habitats to the first four levels of the hierarchy. For example, the code A.1.1 means a stand of broad-leaved woodland which is easy to assess from a brief site-visit. However, detailed field surveys are necessary to define the most refined levels of the habitat hierarchy (levels 5 and 6), where critical differences between woodland types are required e.g. the presence of coppiced (regularly cut) or un-coppiced shrubs. Moreover, definition of the habitat to the level of the vegetation community often requires quantitative measurement of plant species abundance in a number of quadrats, followed by running the data set through a system of keys to assign it to a particular community (e.g. Rodwell 1991).

Similar, although often less precise, systems of habitat classification exist in most other countries world-wide.

By marking the positions of birds on a habitat map the most appropriate level of habitat mapping to elucidate factors of importance to the bird can be determined. In the example given in Box 10.2 the habitat map showing the structural variation in the woodland best explains the distribution of the Nightingale. This bird is a scrub specialist and its song locations (breeding areas) closely follow the distribution of coppice-with-standards woodland, where there is a dense shrub layer. This distribution pattern could be further explained in terms of measured habitat variables as discussed later.

The advantage of using a standard system to record habitats and vegetation communities is that all maps produced will be in the same ecological language and hence inter-site studies will be facilitated. There are several disadvantages, however. These are that botanical specialists may be required to identify sufficient plants to define the vegetation communities, and that a reasonable level of knowledge of ecology and of the site are necessary to define habitats to a fine level. Such studies are also labour-intensive, hence relatively costly.

There are many examples of studies which have made use of habitat mapping techniques to understand bird counts; some of these are presented below.

Examples of the use of habitat mapping studies

1. Satellite mapping of remote sites

Satellite images have recently been used to map habitats in the Flow Country of Caithness and Sutherland in northern Scotland and thereby assess breeding populations of Dunlin (Avery and Haines-Young 1990). In this study coloured Landsat images, produced using the near-infra-red band 7 which is sensitive to vegetation type and ground wetness, were used to map areas of differing habitat. Then by using prior knowledge of the abundance of Dunlin, which suggested they should be most abundant in the wettest areas, total numbers of these birds breeding in a random selection of 2.5×2.5 km squares in the study areas were predicted. These predictions were then tested by field counts of the Dunlin (Chapter 7) in a selection of sites ascribed to different vegetation types. A high level of correlation between the number of Dunlin estimated from the Landsat image and the number counted on the ground was obtained (Box 10.4). One advantage of this method was that it produced data from an extensive area which would have been difficult to survey adequately using other methods. Major disadvantages are the high cost of the satellite photographs, and the fact that these photographs cannot show fine scale of habitat intergradation which are important to many birds. In such cases satellites are not an appropriate method for mapping habitats and assessing bird populations.

2. Habitat mapping from aerial photographs and topographical maps

The habitat preferences of the European Woodcock in Ireland have been assessed using habitat maps. The base habitat map of the study plot was created from an Ordnance Survey map, aerial photographs and ground survey (Wilson 1982). The radio-located positions of 12 Woodcock equipped with radio-transmitters were then marked on the map at dawn and dusk (Box 10.5). By this means, both the nocturnal and diurnal habitats of the Woodcock were assessed, and a habitat preference index was calculated. This study showed that the birds spent the day in the densest area of the forest and ranged more widely over arable and grassland habitats during the night.

Box 10.1   Scale of habitat recording for bird studies.

Box
10.1
cont.

(a) All habitats are mapped without any habitat measurement and the locations of birds are marked on the habitat map (●). This method produces a broad understanding of the birds' habitat preferences, but it is difficult to test any relationships statistically.

(b) Habitat is subdivided into parcels on the basis of criteria such as vegetation age or plant species composition (□ = recent coppice coupes; ▨ = old coppice coupes). Bird registrations (●), derived from a mapping census, are allocated to each parcel and compared with quantitatively measured habitat variables. The habitat data from the parcels are produced independently of the territory mapping census and a statistical comparison between the two to test any significant relationships is possible.

(c) Habitat variables are recorded in standard sample plots at measured distances along the route of a transect bird count. This produces data on habitat variables at the same time and in the same position as the transect count and allows the use of multivariate statistical methods to test relationships between birds and habitat variables. Because transects usually involve walking at a regular speed (Chapter 4) and birds flee from an observer on open ground, this may be a poor method, unless habitat variables can be measured very quickly, or after the birds have been counted. (x) = transect band width, (y) = measured transect segments, (z) = example radius of habitat recording circle.

(d) Habitat variables recorded in sample plots around the position of randomly located point counts. This produces detailed habitat data in the same position and at the same time as the point count. Again this method allows the use of multivariate statistical methods to test relationships between birds and habitat variables. As described in Chapter 5, this method works best in fine-grained habitats such as woods. The relatively poor visibility in woodlands also allows the habitat variables to be measured without disturbing the birds greatly.

(e) Habitat variables are recorded at the position of a territorial, feeding, or radio-located bird. This produces precise habitat data in an area selected by the bird. By also recording habitat variables at a random selection of plots within the study area it is possible to quantify habitat selection by the birds in terms of measured differences in habitat variables that were selected and avoided.

Box
10.2

Scales of habitat mapping in relation to bird distribution (modified from Fuller *et al.* 1989).

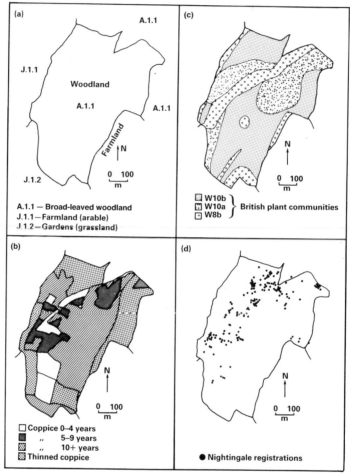

(a) Habitats in the study plot are mapped at the crudest habitat scale. Broad-leaved woodland (code A.1.1) is recognised but this does not predict the distribution of breeding Nightingales presented in (d).

(b) Habitats in the study plot are mapped at a more detailed level where coppiced (cyclical cutting of shrub species) areas are separated by age. The areas of coppice of between 5 and 9 years' growth predict the distribution of Nightingales (d) quite well.

(c) Habitats in the study plot are mapped down to the level of the standard British Plant Community sub-community (Rodwell 1991). Three vegetation types are recognised in the wood, and woodland types W8 and W10a predict the distribution of Nightingales (d) quite well. There may well be a correlation between the vegetation communities presented in (c) and the areas chosen for coppicing presented in (b); this could be investigated further.

(d) Distribution of Nightingale registrations in 1970 at Ham Street Woods, Kent (source Fuller *et al.* 1989).

British habitat classification system to the level of vegetation community for woodlands and scrub (from NCC 1990 and Rodwell 1991).

Box 10.3

First level hierarchy	Second level hierarchy	Third level hierarchy	Fourth level hierarchy
(A) Woodland and scrub	(1) Woodland	(1) Broad-leaved (2) Coniferous (3) Mixed	(1) Semi-natural (2) Plantation

Fifth level hierarchy	Vegetation community
(1) High forest (2) Coppice (3) Coppice-with-standards (4) Orchard (5) Underplanted (6) Oak pasture (7) Unmanaged	One of 25 possible 'British Plant Community' woodland/scrub vegetation communities e.g. *Fraxinus excelsior–Acer campestre* woodland (W8)

Use of satellites to predict bird numbers from habitat data.

Box 10.4

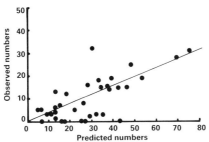

Relationship between Dunlin numbers in northern Scotland predicted from Landsat images and counts obtained by a field survey using standard methods (from Avery and Haines-Young 1990). There is a significant correlation between these two variables. This shows that the method can be used to obtain population estimates for extensive and remote areas without the need for intensive field surveys; only verification surveys are required to check the accuracy of the estimate. The predicted slope is 0.4 (plotted for reference), with an intercept at (0,0). The observed slope of the regression was $+0.39 \pm 0.07$; observed intercept, -1.7 ± 2.12. The technique could be applied to many other species inhabiting relatively simple habitats (e.g. waders on different types of mudflats or mires) during their breeding or non-breeding seasons.

Reprinted by permission from *Nature*, **344**, pp. 860–862. Copyright © 1990 Macmillan Magazines Ltd.

Box
10.5

Habitat use by Woodcock in Ireland.

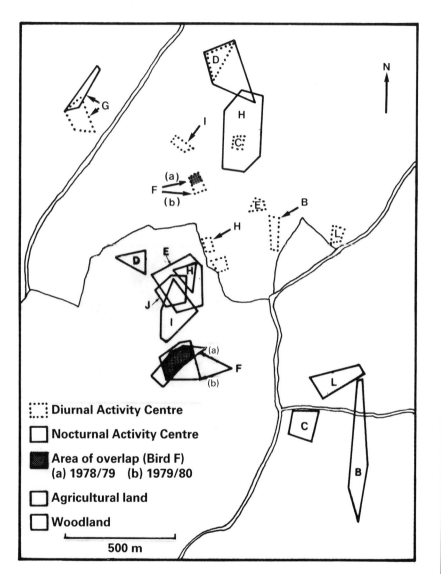

Diurnal Activity Centre

Nocturnal Activity Centre

Area of overlap (Bird F)
(a) 1978/79   (b) 1979/80

Agricultural land

Woodland

500 m

The nocturnal and diurnal distribution of ten radio-tagged Woodcock (labelled A—J) in a study site near Rathdrum in County Wicklow are presented (from Wilson 1982). This shows a clear tendency for these birds to use the woodland habitats, especially the young thicket stages of plantation conifers during the day ('diurnal'), and agricultural land during the night ('nocturnal').

Selection of habitat variables measured in grasslands of North America and related to populations of birds (from Wiens 1969, 1973; Rotenberry and Wiens 1980).

Box 10.6

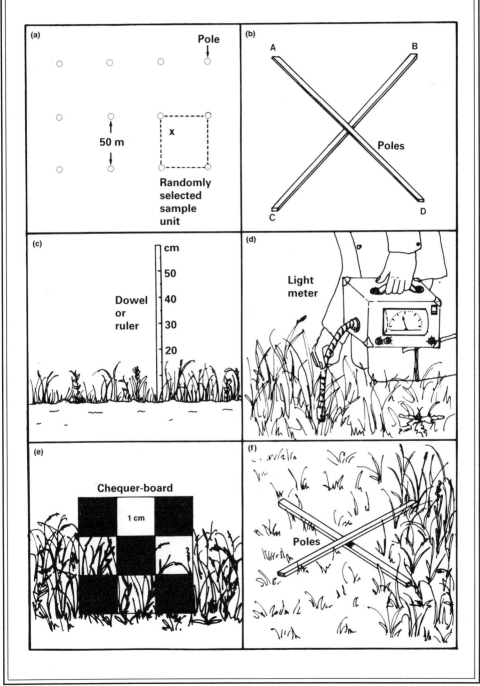

(*Continued*)

Box 10.6 cont.

(a) Habitat variables and bird numbers are measured in 10 ha sample plots subdivided into sampling units at 50 × 50 m—within each unit a randomly located sampling position (marked ×) is used to record the habitat variables.

(b) Wooden poles 2 m long are used to create a cross on the grass—the ends of the cross mark the positions of the four vegetation sampling points (A–D).

(c) A wooden dowel sub-divided into 10 cm units is placed vertically into the grassland and used to measure the depth of the grass and the litter.

(d) A portable light meter (photometer) can be used to measure light penetrating into the grassland to produce an index of vegetation density at different heights.

(e) A chequered board with 5 × 5 cm subdivisions is used to produce an index of vegetation density at various heights.

(f) The patchiness in the height of the vegetation at the various corners of the poles can be used to calculate a heterogeneity index for the vegetation.

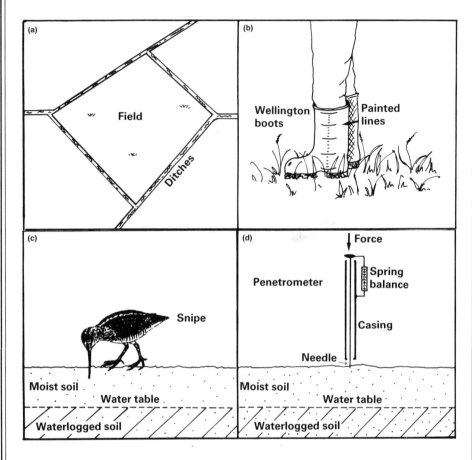

Selection of habitat variables measured in lowland wet grasslands of Britain and related to bird populations (from Green 1985b, 1988).

(a) The study plot is defined by the boundaries of an individual field surrounded by ditches.

(b) By marking height lines on the wellington boots worn by the observer the height and heterogeneity of the vegetation can be rapidly assessed.

(c) Many species of wader breeding in lowland wet grasslands in Britain (particularly Snipe) need soft soils in which to probe for invertebrate food. The important feature is that the water table is close to the surface of the soil during the breeding season and this makes the soil soft.

(d) Penetrometers simulate a bird's beak and can measure the penetrability of the soil in terms of the force needed to probe it. The value produced is an index of the soil conditions required for Snipe and other probing birds to obtain food.

**Box
10.8**

Wader nesting density in relation to vegetation height.

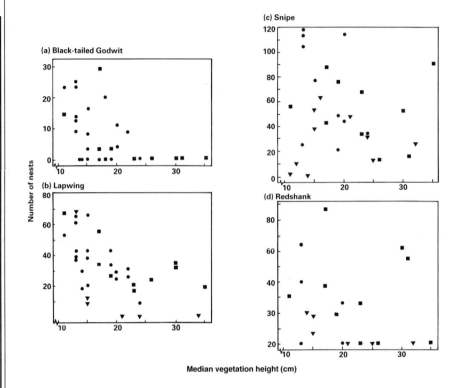

Density of nests of (a) Black-tailed Godwit, (b) Lapwing, (c) Snipe and (d) Redshank with respect to median vegetation height in mid May at three lowland wet grasslands in southern Britain (from Green 1985b).

● = data from the Ouse Washes (Cambridgeshire), ▼ = data from the Nene Washes (Cambridgeshire), ■ = data from West Sedgemoor (Somerset and Avon).

Examples (a) and (b) show that Black-tailed Godwit and Lapwing are most common in fields with a median vegetation height of 10–20 cm in May, whereas examples (c) and (d) show that breeding densities of Redshank and Snipe are less influenced by the recorded vegetation heights.

	Box
Commonly recorded grassland habitat variables and examples of methods used to measure them.	**Box 10.9**

Variable	Methods of recording
Vegetation height	Wellington boot marked into bands of 5 or 10 cm Stick marked into bands 1–10 cm high
Vegetation density	Chequered board marked with height bands and height of 90% obscuration of bands read at 5 m distance Light meter lowered a known distance into the vegetation and light intensity measured
Vegetation heterogeneity	Variability in vegetation height calculated for 50 readings over e.g. a field
Litter depth	Ruler used to measure litter depth directly, averaged over a representative number of samples (at least 30)
Grazing regime	Presence and type of grazing animals Number of livestock grazing-units per ha per year
Plant species diversity	Number of plant species in representative (at least 20) quadrats of 1 × 1 m or 2 × 2 m
Vegetation community	Assignment of vegetation to standard vegetation community based on data collected in at least five standard quadrats
Soil softness	Measurement of penetrability of the soil using a mechanical device which measures force to insert steel rod (penetrometer)
Soil type	Observation in shallow soil pit
Environmental factors	Rainfall, temperature, altitude, latitude/longitude, season etc.
Natural grassland or Ley	Historical knowledge, presence of indicator species

Box 10.10

Some commonly used devices to measure habitat variables in woodlands.

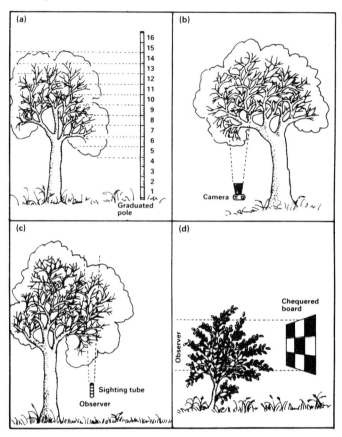

These devices have most frequently been used to assess the vertical density of foliage in a forest to produce a foliage profile, but they can also be used to generate independent data on habitat variables to test against the abundances of the various bird species present. Most ideas originate from James and Shugart (1970) and MacArthur and MacArthur (1961).

(a) Graduated pole held upright—most useful to measure the features of the foliage in the shrub layer, and low forests.

(b) 35 mm camera with 135 mm or 'zoom' lens—can be focused down through the forest profile (heights read off range-finder) and used to assess foliage density through a vertical section of the forest.

(c) Sighting tube—observer looks directly up and assesses the canopy or shrub layer foliage density, or attempts to divide the profile into height bands and assesses vegetation cover within each.

(d) Chequered board—used to assess vertical density of shrub layer. Observer walks away from the board until 50% of the board is assessed to be obscured by vegetation; this produces an index of the shrub density which can be repeated at a variety of heights. It is important that the same observer assesses when 50% of the board has become covered as observers may vary in this ability.

Foliage profiles for two hypothetical woodlands and an example of their relationship to bird populations.

Box 10.11

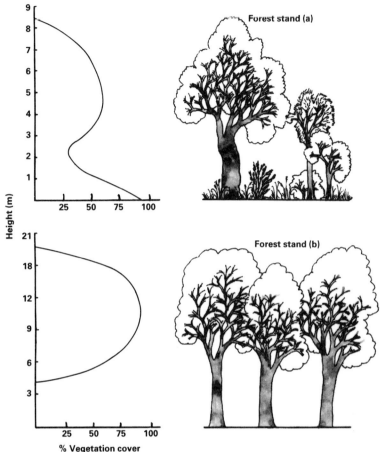

A vertical section of the forest vegetation has been broken into height bands and the foliage density has been either estimated or measured within each band; this information can be used to produce a foliage profile which can be used to calculate a foliage height diversity index, or put to other comparative purposes.

Forest stand (a) is a semi-natural woodland and has a well developed canopy, shrub layer and ground layer—these give rise to a smooth foliage profile. This profile might be expected to favour both species that require height diversity, and also to support a high total diversity of species as there are many ecological niches.

Forest stand (b) is a plantation where all the trees are of uniform age and there is a dense canopy, but virtually no shrub or ground layers. This profile might be expected to support a less diverse bird community than profile (a) as those species that use shrub and ground layers will be absent or very scarce, and species that range widely through different vegetation levels will also be rare. However, some species prefer poorly developed understorey conditions and these might well be more abundant in this woodland type.

Box 10.12	Commonly recorded woodland/scrub habitat variables and examples of methods used to measure them.

Variable	Methods of recording
Canopy height	By trigonometry, with a camera or hypsometer, or directly (at least 20 readings)
Canopy cover	Estimate through sighting tube, or through camera with acetate grid on view-finder (at least 20 readings)
Canopy heterogeneity	Summed data from at least 50 readings on canopy cover analysed by a heterogeneity index
Vertical foliage height diversity	Percentage vegetation cover (to 5%) at various height bands taken vertically through woodland
Horizontal foliage diversity	Variation in cover in various height bands laterally. Can be used to create heterogeneity index
Dead wood	Estimate through sighting tube percentage quantity dead wood in the canopy or on the ground (at least 20 readings)
Ground cover	Estimate herb, leaf-litter, twig, moss cover using e.g. a quadrat of 0.5 m^2 or sighting tube (at least 20 readings)
Shrub density at various heights	Use chequered board to produce at least 20 half-sighting distances at various heights in the shrub layer. Normal heights are 0.5, 1.0 and 1.5 m above the ground
Tree diameter	Diameter of at least 20 trees at breast height
Tree age	Knowledge of planting regimes, cores through tree-trunks, or simply diameter
Broad-leaved or conifer	Direct observation of stand, percentage frequency occurrence along at least 20 representative transects
Plant species diversity	Assessment of tree, shrub, and ground layer plant species diversity in at least 20 quadrats of 20 × 20 m for trees and shrubs and 5 × 5 m for ground layer
Vegetation community	Assignment of vegetation to standard vegetation community by recording species composition and relating to reference documents
Natural forest or plantation	Historical knowledge, presence of indicator species
Grazing regime	Presence and type of grazing animals Number of livestock-units per ha per year
Soil type	Observation in shallow soil pit Geological/soil survey maps
Environmental factors	Rainfall, temperature, altitude, latitude/longitude season, etc.

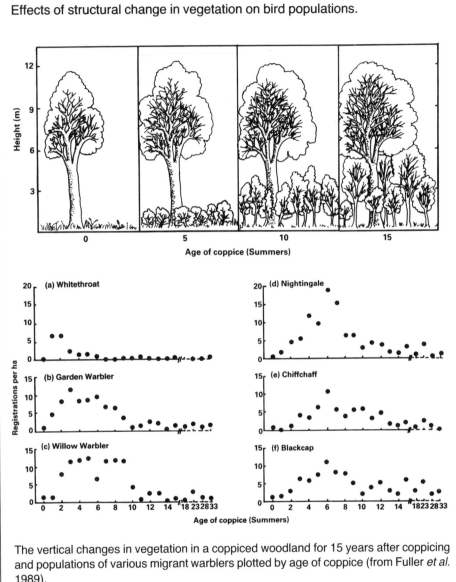

Effects of structural change in vegetation on bird populations.

Box 10.13

The vertical changes in vegetation in a coppiced woodland for 15 years after coppicing and populations of various migrant warblers plotted by age of coppice (from Fuller *et al.* 1989).

Following coppicing the cut shrubs grow rapidly and there is a rapid change in the vertical vegetation structure and shrub density of the example woodland. This rapid structural change is reflected in populations of various African migrant warbler species (a–f) which select different structural characteristics (in this case age/density) of coppice. After around 10 years of growth the structure of the coppiced shrub layer is changing much more slowly and the bird community becomes more stable. Populations of the African migrant warbler species fall, and are replaced by resident birds such as tits and thrushes (not shown). The species graphs are presented (a–f) approximately in order of habitat selection with species preferring the youngest coppice coming first.

| Box 10.14 | Habitat factors making a significant ($P < 0.05$) contribution to explaining bird numbers, as assessed by multiple regression analysis (from Bibby and Robins 1985). |

	Variables													% Variance explained
	1	2	3	4	5	6	7	8	9	10	11	12	13	
Blackcap							+							54.7
Willow Warbler		+	+	+										49.8
Chiffchaff			+											45.7
Wood Warbler				+	+						−			71.0
Goldcrest		−		−										62.9
Pied Flycatcher	−								+	+	−			72.5
Redstart							−	−						59.3
Blackbird		+	+		+									64.3
Willow Tit									−		−		+	60.6
Nuthatch							I					−	+	57.5
Treecreeper	−						−							54.0
Wren												+		61.6
Song Thrush														NS
Robin														NS
Great Tit														NS
Blue Tit														NS
Chaffinch														NS

Key to variables: 1 = birch %, 2 = sessile oak %, 3 = tree species diversity, 4 = canopy cover diversity, 5 = shrub cover, 6 = foliage height diversity, 7 = herb cover, 8 = density trees > 15 m, 9 = % overmature trees, 10 = holes, 11 = hazel %, 12 = bramble %, 13 = scrub height diversity.

The higher the variance explained the better the correlation with that particular variable. Variations of over 60% are highly significant.

Radio-telemetry as a tool for studying habitat preferences of individual species.	**Box 10.15**

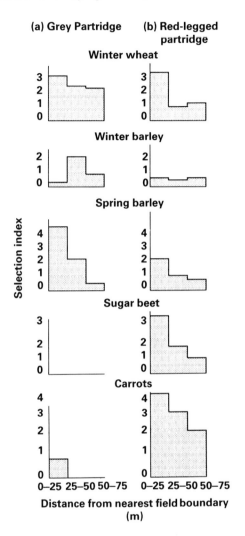

Habitat selection by (a) Grey Partridge and (b) Red-legged Partridge in Britain is shown (from Green 1984). Histograms show ratios of observed to expected numbers of radio-locations in five crops and at various distances from the nearest field boundary. This shows the preference of these two species for areas adjacent to hedges, and also shows that they prefer different crop types. But see page 203.

Box 10.16	Means and standard deviations of habitat variables that were significantly different ($P < 0.05$) between Woodcock feeding locations as determined by radio-telemetry of marked birds and random sites (Hirons and Johnson 1987).

Variable	Woodcock feeding areas Mean SD		Randomly chosen areas Mean SD		Significance level
Vegetation structure					
Basal area of trees (m² per ha)	2.8	1.73	4.4	4.20	<0.05
Mean basal area of trees (cm²)	27.8	10.50	37.7	18.50	<0.01
Height of co-dominant vegetation (cm)	8.2	6.96	4.9	6.73	<0.05
Vegetation composition					
Dog's Mercury (% cover)	19.9	13.34	1.3	5.37	<0.001
Beech (% of point quarters)	12.4	25.30	40.6	40.63	<0.05
Oak (% of point quarters)	3.2	7.78	0.3	3.08	<0.05
Ground surface and soil characteristics					
Litter (% cover)	41.5	14.52	62.6	25.00	<0.05
Litter depth (cm)	1.5	1.14	2.5	2.67	<0.05
pH	6.3	0.79	5.2	1.15	<0.001
Earthworm biomass (g per 0.25 m²)	5.62	3.21	3.08	3.27	<0.001
Earthworm numbers (per 0.25 m²)	18.92	1.74	10.41	1.80	<0.001

In this analysis the lower the level of P the higher the significance of the result, hence the most significant results are:

(1) the preference for feeding within Dog's Mercury;

(2) the preference for feeding in areas with higher soil pH, earthworm numbers and earthworm biomass.

Measuring habitat variables in sample plots

Mapping procedures are able to provide broad information on the habitat preferences of a particular bird, or assemblage of birds, but cannot define which features of the habitat are of most importance.

Plot-based studies generally aim to refine knowledge on those habitat variables that are most important to the birds. Because these methods may collect large volumes of data, interpretation of the bird-count and habitat-variable data necessitates analysis with sophisticated multivariate statistical procedures (e.g. Gauch 1982). For example, the density of the shrub layer might be expected to be an important factor in determining the distribution of a shrub-dwelling woodland bird; this could be measured at the same time as counts of the bird were made (see Chapters 4 and 5) and possible relationships tested.

Several methods have been developed to measure habitat variables in grassland and forest habitats and then relate these to a bird's distribution and abundance. Practical details of the methods of recording habitat variables in grassland and woodland/scrub habitats in relation to counts of birds are presented below.

Grassland habitat variables

Study plots in grasslands should be large enough to gain an adequate sample of the bird species being studied, but not so large that vegetation features change dramatically within the plot. Moreover, enough plots should be studied to enable statistically meaningful results to be obtained; over 20 study plots are generally recommended. Plots are normally positioned within different grassland management regimes such as grazing or mowing (stratified random sampling), as these will be important when the results are being analysed. Within these constraints, study plots should be randomly located, and vegetation sampling positions should also be randomly located within the plots.

The distribution and abundance of grassland birds appears to be most influenced by habitat variables such as vegetation height, vegetation density and heterogeneity, vegetation composition, grazing density, soil-moisture content and ease of obtaining food. These variables are generally those sampled in ornithological studies.

Examples of studies using grassland habitat variables collected in study plots

1. Habitat preferences of breeding birds of North American prairies counted using mapping methods

In North America, the effects of vegetation height, vertical vegetation density, vegetation heterogeneity, depth of the litter layer and grazing pressure on the breeding abundance of selected grassland birds have been intensively studied (Wiens 1969, 1973; Rotenberry and Wiens 1980).

In these studies, habitat and bird data were collected in 10-ha study plots in representative areas of vegetation throughout the mid-west of North America. Each study plot was demarcated by a grid marked out with stakes around 60 m apart. All plots were surrounded by a 'buffer zone' of similar vegetation at least 100 m wide in order to minimise edge effects. Breeding densities of birds in these plots were assessed using a mapping method with the birds being flushed to define their territory boundaries accurately (consecutive-flush method: see Chapter 3).

Features of the vegetation structure in these plots were recorded at sampling units located randomly within each 61 × 61 m block (Box 10.6a). At
each sampling unit, four sampling plots were located on the ends of 2 m long
poles arranged in a cross (Box 10.6b). The following vegetational attributes
were recorded.

(1) Vertical vegetational structure: this was recorded by noting the number of vegetation contacts along a 1 m long wooden dowel of c. 5 mm
thickness sub-divided into 10 cm bands positioned vertically in the vegetation (Box 10.6c). The type of vegetation making contact with the dowel
and the depth of litter or mulch could also be recorded.

(2) Vegetation density: this was assessed as the quantity of light, as
measured by a portable field photometer, penetrating a known distance into
the vegetation (Box 10.6d). Readings were taken at the same time of day and
under similar weather conditions to produce standardised results. Another
means of measuring vegetation density involved placing a 10 cm wide board,
marked into 1 × 1 cm squares, vertically into the vegetation (Box 10.6e).
From a distance of 5 m an estimate was made of the height on the board
where 90% of the squares were obscured by vegetation.

(3) Vegetation heterogeneity: the height of the vegetation at the four
corners of the 2 m square quadrat were measured (Box 10.6f) to enable
vegetation heterogeneity to be calculated, as below:

$$\text{Heterogeneity index} = \frac{\Sigma(max - min)}{\Sigma x}$$

where *max* = maximum height of the vegetation in the quadrat, *min* = minimum height of the vegetation in the quadrat, and *x* = mean height of the
vegetation in the quadrat. Low values of this index indicate uniformity of the
vegetation and high values heterogeneous vegetation. Values can be summed
over the whole study plot to give an overall heterogeneity measure which can
be compared with other plots and against the different abundance of birds.

2. Habitat preferences of probing wetland waders in Great Britain counted using specialised counting methods

Research into the habitat preferences of breeding waders on the lowland wet
grasslands of Britain (Green 1985a, b, 1988) also shows how measured
habitat variables in sample plots can be used to understand the abundance of
breeding birds. In these studies the sample plots were defined by field
boundaries. At each site populations of various breeding wader species were
assessed in several fields using standard methods (Chapter 7). Habitat
variables were also measured using the following methods.

(1) Structural variation of the sward: this was sampled every 20 paces
along randomly orientated transects through fields. Vegetation height was
measured next to the observer's foot, at a rate of around 10–20 sample points
per ha of field. By painting white height lines on the wellington boots worn
by the observer data collection was speeded up considerably, without loss of
accuracy (Box 10.7b).

(2) The ease of penetration of the soil: this controls whether the birds can continue to probe the soil and hence obtain invertebrate food. Penetrability was measured with a penetrometer, a device comprising a metal needle which mimics a wader's beak, attached by a linkage system to a 10 kg balance. At each sampling point (where vegetation height was measured) five penetration measurements were made by sticking the needle 10 cm into the ground. From 30 to 60 penetration readings were collected per km^2 of study area (Box 10.7d).

Data collected were used to demonstrate the relative importance of these habitat factors to a variety of waders breeding in lowland wet grasslands. For example, by presenting the vegetation height data as the percentage falling within various height classes it was shown that the highest density of nesting Lapwing and Black-tailed Godwit occurred in fields with short grass up to 10–20 cm median vegetation height in mid May (Box 10.8). In comparison, Snipe and Redshank nested in grasslands over a wider range of vegetation heights (Box 10.8).

A more complete list of the variables that can be recorded in a grassland with the aim of investigating bird/habitat relationships is presented in Box 10.9.

Woodlands and scrub

Woodlands are complicated 3-dimensional habitats, so describing their variations at a scale applicable to bird studies is more difficult than for grasslands. Many methods have been used in woodlands and there has been little standardisation of what has been measured and the ways in which this has been done. In this section, methods are presented to measure some of the most commonly recorded woodland habitat parameters.

As with grasslands, habitat variables of probable importance to an individual species or assemblage of woodland or scrub birds can be collected in sample plots. Plots can be positioned (1) at the site of a randomly located point count (Chapter 5), (2) at regular intervals along transects (Chapter 4), (3) in relation to the distribution of mapped bird territories (Chapter 3), or (4) at the position of a singing or radio-located bird.

1. Circular sample-plot method

The 'circular sample-plot' is widely used to collect data on habitat variables in woodlands and scrub (James and Shugart 1970). The standard sample-plot is circular and 0.05 ha in extent (12.62 m radius). Within this plot a number of habitat variables can be quantitatively recorded or estimated depending on the time available for the study and the precision of results being sought. In cases where several observers are being used to collect habitat data it is important that they all receive a day's training in order to minimise observer variation.

2. Habitat variables measured in woodlands and scrub

Methods of measuring the habitat variables most commonly recorded in studies of woodland and scrub birds, because ecological knowledge suggests they are most important to these species, are described below and in Box 10.10.

(1) Tree species and diameter: each tree is identified to species level and the diameter at breast height recorded either to the nearest cm, or more commonly within bands of 5 or 10 cm. Tree diameter gives an index of forest maturity, especially if the same tree species are compared between otherwise similar forests.

(2) Presence/absence of dead wood: the quantity of dead wood on the forest-floor or in the canopy can provide both an index of its availability and some idea of forest maturity as older woodlands generally have larger quantities of dead wood.

(3) Tree and shrub diversity: the number of species of trees and shrubs can be recorded within the whole plot.

(4) Ground cover: the percentage cover of ground vegetation, leaf-litter, sticks or bare ground can be evaluated using a 0.5×0.5 m or 1×1 m quadrat. An index of ground cover by vegetation can also be calculated from a number (e.g. 20) of plus or minus readings made through a sighting tube (plastic or metal tube) pointed directly down at the ground. Suitable tubes are between 2 and 5 cm in diameter.

(5) Canopy cover: canopy cover can be assessed as a percentage through a zoom lens attached to a camera, or a sighting tube. An index of the estimate of the canopy cover can also be made by taking a number (e.g. 20) of plus or minus readings for the presence or absence of green leaves sighted directly upwards on alternate steps along a transect through the circle.

(6) Canopy height: the average height of the canopy can be measured with a measuring device such as a clinometer, by using trigonometry, or read off the range-finder scale (logarithmic) of a camera lens held vertically and focused on the top of the canopy.

(7) Shrub density: the density of the shrub layer foliage can be calculated by means of a standard 30×50 cm board which has been painted with red (or black) and white squares of 10×10 cm (see Fuller *et al.* 1989). When the chequered board is being used one observer holds the board at a predetermined height in the vegetation and the other walks away until the board is half obscured by vegetation. This 'half-sighting distance' is then measured to give an index of vegetation density whereby the shorter the sighting distance the denser the vegetation. These boards are normally used to record shrub densities at three different heights in order to assess height variations. Commonly chosen heights in British woodlands are 0.5 m, 1.0 m and 1.5 m. However, the height of recording is often varied according to local conditions. Between 10 and 50 half-sighting readings are commonly made in each woodland study site, with the number of readings being controlled by the variability of the results—the more variability the more replicates will be needed. The density of shrubs near the ground can also be recorded along

two transects of outstretched armlength width (c. 2 m) across the circle, each totalling approximately 0.02 acres (0.008 ha). All contacts with shrubs along these transects are noted, in different size categories if appropriate.

(8) Vertical vegetation density: the density of the canopy, shrub layer and ground layer can be measured separately as defined above, or regarded as components of a vertical vegetation profile through the forest. Such 'foliage profiles' are calculated by dividing a vertical section of the forest into defined height bands and then assessing the foliage cover within each band (Erdelen 1984; Petty and Avery 1990). Vegetation density can be measured in several ways.

(a) Estimation: the most rapid technique to produce a foliage profile involves assessing the maximum and minimum foliage heights, either by visual estimation or with some form of measuring device, and then estimating the percentage vegetation cover to the nearest 5% at various heights through the profile. A profile of each study site can be sketched in the field using the upper and lower canopy and the maximum percentage cover as guide-lines.

(b) Sighting tube: the vertical density of vegetation can also be measured through sighting tubes 5–20 cm in diameter. These are pointed directly up into the forest and by looking through them and focusing by eye through the forest layers (Box 10.10c) a visual estimation of the foliage densities can be produced.

(c) Graduated pole: another rapid method to assess the foliage profile uses a long thin rod graduated along its length. This pole is positioned vertically through the forest profile and the number of foliage contacts within measured sections of the pole are recorded. These contacts can be converted to a foliage density at the various heights and this can be used to produce a foliage profile. This method is most successful in low-growing canopies as positioning and reading the pole becomes increasingly difficult in taller stands (Box 10.10a).

(d) 35 mm camera: vertical sightings are taken through a 135 mm or 'zoom' lens attached to a standard 35 mm single lens reflex camera. The density of intersected leaves can be estimated at various heights defined by the range-finder scale on the camera lens (Box 10.10b). By using a grid marked on an acetate film placed over the camera eyepiece more detailed quantification of the vegetation cover at different heights can be obtained. Foliage profiles can be plotted from these data. 35 mm cameras with 'fish-eye' lenses can also be used as they produce a photograph of the whole of the profile, and the foliage density of the various layers can be measured off the photographs.

(e) Chequered boards can also be used to measure the density of vegetation at various heights (Box 10.10d), although this becomes increasingly difficult further from the ground.

Examples of profiles from two contrasting hypothetical woodlands are presented in Box 10.11. Classic research in the early 1960s indicated that the diversity of bird species was directly related to the diversity of vegetation in the foliage profile (MacArthur and MacArthur 1961; MacArthur *et al.* 1962). This has been elaborated many times since then (e.g. Wiens 1989).

To test this relationship, foliage profiles must be turned into foliage height diversity values using the Shannon–Weiner formula:

$$H = -\Sigma \, p_i \ln p_i$$

where p_i is the proportion of the total foliage which lies in the ith of the chosen horizontal layers.

Thus, for instance, a woodland with one layer has zero diversity. Two layers, one with 1% cover and the other with 99% cover will have a diversity of $-0.01 \ln 0.01 - 0.99 \ln 0.99 = 0.056$ (close to zero), while two layers each with 50% cover will have a diversity of $2 \times (-0.5 \ln 0.5) = 0.694$. This illustrates why **FHD** is a better measure of diversity than the actual number of layers, for the community with 99 of one species and one of the other seems closer to the community with only one species.

FHD values so calculated can then be compared with bird species diversity produced using the same formula. As mentioned above bird species diversity and foliage height diversity have been positively correlated in many studies.

As many further habitat variables and physical/climatic features of the sample areas as considered appropriate can be measured in each sample plot. Box 10.12 summarises sorts of habitat variables that can be measured to gain useful knowledge on the habitat preferences of birds.

Examples of the use of quantitative measurement of structural variables in study plots in woodlands and scrub

1. Mapping method used to study the effects of shrub layer density on populations of breeding birds

Historically, many British woods were actively coppiced (shrub species cut down every 7–15 years in a regular cycle). Within each coppiced area there are extensive changes in the vertical vegetation structure as the coppice regrows (Box 10.13). The detailed effects of this vegetation change on populations of various breeding birds have been investigated by Fuller *et al.* (1989). Positions of all bird registrations were mapped in Ham Street Woods, Kent over a 5-year period with 23–25 census visits annually. Vertical vegetation structure was recorded using chequered boards at 0.5 m and 1.5 m above the ground in stands of known age from the last coppicing. This work showed that coppiced shrubs in the woodland grew quickly and reached their highest vegetation density after 3–5 years, thereafter declining (Box 10.13). Different species of scrub-specialist migrant warblers were found at maximum densities in coppice of different age, particularly in the first 2–10 years following coppicing when the density of the shrub layer was changing most rapidly.

2. Point counts used to assess effects of vegetation structure on bird populations of sessile oakwoods in western Britain

The point count (Chapter 5) method of counting birds has the advantage over mapping methods (Chapter 3) of being able to collect habitat data in a sample plot centred on the point count immediately following the bird count. Also, only one site visit is necessary to collect both bird and habitat data whereas around ten visits are necessary with the mapping methods. As a consequence, point counts of birds allied with habitat measurement allow data to be rapidly collected from many sites and permit detailed statistical investigations.

For example, by estimating 13 habitat variables at the position of randomly located point counts in sessile oak woods in the west of Britain, Bibby and Robins (1985) were able to investigate statistical relationships between birds and their habitat. Box 10.14 gives the results of a multiple regression analysis on the bird counts and habitat variables.

This study provided clues on the most important habitat features for these various birds. For example it suggests that of the measured habitat variables, the presence of a herb layer was most correlated with numbers of Blackcap. Redstarts by contrast showed a negative correlation with herb cover, reflecting the fact that this species prefers more open, often heavily grazed, woodlands. The numbers of Robin, Great Tit, Blue Tit, Song Thrush and Chaffinch were not statistically correlated with any of the measured habitat variables, probably because these species are quite flexible in their habitit requirements. Such information is invaluable when conservation action is being planned, or habitat management programmes are being designed.

Similar studies can be undertaken using transect counts of birds with habitat measured at either regular, randomised or selected positions along the transect (e.g. Hill *et al.* 1990, 1991).

It is important to note that spurious correlations may result when habitat variables are themselves correlated. For example, a dense shrub layer will tend to produce a sparse ground layer and both of these may be statistically significantly correlated with the numbers of a particular bird, although a broad ecological understanding of the bird would suggest that it spends all its time in the shrub layer and never uses the ground. However, on occasion apparently spurious correlations may prompt a useful re-investigation into the habitat preferences of the bird!

Individual-based studies

As well as relating bird counts to mapped habitat features, and habitat variables measured in sample plots, it is also possible to record habitat variables at the exact location of a singing bird, or one located by radio-telemetry. Such studies can be used to define habitat selection by a bird more precisely, and if habitat variables are recorded at positions with and without a bird, then a habitat preference index can be created (see later).

Examples of the use of measuring habitat variables at the position of an individual bird

1. Radio-marked partridges on farmlands in Britain

An example of recording habitat features in grasslands at the precise location of a study bird is provided by Green (1984). In this investigation the early morning (roosting site) locations of female Red-legged and Grey Partridges with young were assessed by radio-telemetry. Then by visiting the precise location of the radio-fix and finding the roosting area, habitat variables such as crop type and distance from the nearest hedge were measured. Moreover, droppings could be collected and used to assess what the birds had been feeding on.

   After many such sites had been visited and habitat variables collected and analysed, it could be shown that both species roost most often close to hedges, but that the habitat preference varies between the species. Grey Partridge preferred cereals, whereas Red-legged Partridge preferred carrots and sugar beet crops within the study area (Box 10.15).

2. Radio-marked Woodcock feeding in woodlands and fields

The study of Hirons and Johnson (1987) recorded details of woodland habitat in a study area, and used radio-telemetry to locate Woodcock within it and assess their habitat preferences. To describe the habitat the study plot was first divided into four strata (trees, saplings, shrubs and herbaceous). Radio-tagged Woodcock were then located within these broad habitat divisions by radio-telemetry and details of the habitat were collected by means of a 0.25 m^2 quadrat at feeding, nesting and randomly located sites. As an example, 30 habitat variables were recorded at each of the 50 feeding locations. By recording habitat variables in areas used by feeding Woodcock and in randomly located quadrats a habitat preference index could be developed, showing that the Woodcock positively selected some of the habitat attributes in the study area (Box 10.16). For further information on the use of radio-telemetry data see Chapter 9.

Summary and points to consider

Bird habitats can be measured at various intensities from mapping the habitat and marking the positions of birds, to measuring habitat variables at the position of a study bird. Habitats are generally recorded within three broad divisions: (1) mapping methods—producing habitat maps; (2) sample plot methods—measuring habitat variables in representative sample plots; (3) individual-based methods—measuring habitat variables at the known position of a study bird.

1. Mapping methods — habitat maps

Base habitat maps are produced from national geographical maps, aerial photographs or satellite images. They are refined by ground survey and, at the highest intensity of effort, habitats can be classified down to the vegetation community level. If registrations of birds are marked on habitat maps they provide basic information on the birds' habitat preferences, but cannot determine which detailed features of the habitat are most important to the birds. Given prior knowledge of the habitat-preferences of a particular bird species, habitat maps can be used to predict bird distribution and population levels over extensive areas.

2. Measurement of habitat variables in sample plots

Habitat variables can be recorded in sample plots at the position of a point count or along a transect. Measurement of habitat variables of probable importance to the bird, as derived from prior knowledge of its ecology, enables the relative importance of habitat features to be investigated. It is then possible to discover which habitat variables are the most important to an assemblage of birds, or an individual bird species and act on this knowledge.

3. Measurement of habitat variables at the position of a study bird

Habitat variables can also be measured at the precise position of a breeding or feeding bird (e.g. located by radio-telemetry). A more refined level of knowledge of the important habitat features to the bird should result from this form of study.

Appendix

Scientific Names of Species Mentioned in the Text

Mammals

Fox   *Vulpes vulpes*

Plants

Ash   *Fraxinus excelsior*
Beech   *Fagus sylvatica*
Birch   *Betula* spp.
Bramble   *Rubus fruticosus* agg.
Dog's Mercury   *Mercurialis perennis*
Field Maple   *Acer campestre*
Hazel   *Corylus avellana*
Pedunculate Oak   *Quercus robur*
Sessile Oak   *Quercus petraea*

References

Aebischer, N.J. and Robertson, P.A. (1992). Practical aspects of compositional analysis as applied to Pheasant habitat utilisation. In: Priede, I.G. and Swift, S.M. (eds.). *Wildlife Telemetry: Remote Monitoring and Tracking of Animals*. Ellis Harwood, Chichester, UK.

Aitchison, J. (1986). *The Statistical Analysis of Compositional Data*. Chapman and Hall, London.

Alexander, H.G. (1935). A chart of bird song. *British Birds* **29**, 190–198.

Alldredge, J.R. and Ratti, J.T. (1986). Comparison of some statistical techniques for analysis of resource selection. *Journal of Wildlife Management* **50**, 157–165.

Andreev, A. (1988). The ten year cycle of the Willow Grouse of lower Kolyma. *Oecologia* **76**, 261–267.

Arbib, R.S. (1981). The Christmas Bird Count: constructing an 'ideal model'. *Studies in Avian Biology* **6**, 30–33.

Avery, M.I. (1989). The effects of upland afforestation on some birds of the adjacent moorlands. *Journal of Applied Ecology* **26**, 957–967.

Avery, M.I. and Haines-Young, R.H. (1990). Population estimates derived from remotely-sensed imagery for *Calidris alpina* in the Flow Country of Caithness and Sutherland. *Nature* **344**, 860–862.

Baillie, S.R. (1990). Integrated population monitoring of breeding birds in Britain and Ireland. *Ibis* **132**, 151–166.

Barnes, R.F.W. (1987). Long-term declines of Red Grouse in Scotland. *Journal of Applied Ecology* **24**, 735–741.

Barrett, J. and Barrett, C. (1984). Aspects of censusing breeding Lapwings. *Wader Study Group Bulletin* **42**, 45–47.

Bart, J. and Klosiewski, S.P. (1989). Use of presence–absence to measure changes in avian density. *Journal of Wildlife Management* **53**, 847–852.

Bayliss, P. (1989). Population dynamics of Magpie Geese in relation to rainfall and density: implications for harvest models in a fluctuating environment. *Journal of Applied Ecology* **26**, 913–924.

Begon, M. and Mortimer, M. (1986). *Population Ecology. A Unified Study of Animals and Plants*. Blackwell Scientific Publications, Oxford.

Bellrose, F.C. (1976). *Ducks, Geese and Swans of North America (2nd edn)*. Stackpole Books, Harrisburg.

Bergan, J.F. and Smith, L.M. (1989). Differential habitat use by diving ducks wintering in South Carolina. *Journal of Wildlife Management* **53**, 1117–1126.

Bergan, J.F., Smith, L.M. and Mayer, J.J. (1989). Time–activity budgets of diving ducks wintering in South Carolina. *Journal of Wildlife Management* **53**, 769–776.

Berthold, P., Fliege, G., Querner, U. and Winkler, H. (1986). Die Bestandsentwicklung von Kleinvögeln in Mitteleuropa: Analyse von Fangzahlen. *Journal für Ornithologie* **127**, 377–439.

Bibby, C.J. (1973). The Red-backed Shrike: a vanishing British species. *Bird Study* **20**, 103–110.

Bibby, C.J. (1978). A heathland bird census. *Bird Study* **25**, 87–96.

Bibby, C.J. (1989). A survey of breeding Wood Warblers *Phylloscopus sibilatrix* in Britain, 1984–1985. *Bird Study* **36**, 56–72.

Bibby, C.J. and Buckland, S.T. (1987). Bias of bird census results due to detectability varying with habitat. *Acta Oecologica–Oecologica Generalis* **8**, 103–112.

Bibby, C.J. and Charlton, T.D. (1991). Observations on the San Miguel Bullfinch. *Açoreana* **7**, 297–304.

Bibby, C.J. and Hill, D.A. (1987). Status of the Fuerteventura Stonechat *Saxicola dacotiae*. *Ibis* **129**, 491–498.

Bibby, C.J., Phillips, B.N. and Seddon, A.J. (1985). Birds of restocked conifer plantations in Wales. *Journal of Applied Ecology* **22**, 619–633.

Bibby, C.J. and Robins, M. (1985). An exploratory analysis of species and community relationships with habitat in western oak woods, pp. 255–265. In: Taylor, K., Fuller, R.J. and Lack, P.C. (eds.). *Bird Census and Atlas Studies. Proceedings, VIII International Conference on Bird Census and Atlas Work*. BTO, Tring, Herts.

Bibby, C.J. and Tubbs, C.R. (1975). Status and conservation of the Dartford Warbler in England. *British Birds* **68**, 177–195.

Birkhead, T.R. and Nettleship, D.N. (1980). Census methods for Murres *Uria* species: a unified approach. *Occasional Papers of the Canadian Wildlife Service* **no. 43**.

Bock, C.E. and Root, T.L. (1981). The Christmas Bird Count and avian ecology. *Studies in Avian Biology* **6**, 17–23.

Briggs, K.T., Tyler, W.B. and Lewis, D.B. (1985). Comparison of ship and aerial surveys of birds at sea. *Journal of Wildlife Management* **49**, 405–411.

British Birds (1984). *The 'British Birds' List of Birds of the Western Palearctic*. British Birds, Bedford.

Brown, L.H., Fry, C.H., Keith, S., Newnam, K. and Urban, E.K. (eds.) (1982 onwards). *The Birds of Africa. Vols I–III*. Academic Press, London.

BTO (1984). *Ringers Manual*. BTO, Tring, Herts.

BTO (1989). *Instruction to Counters: Breeding Waders of Wet Grasslands Survey*. BTO, Tring, Herts.

Buckland, S.T. (1984). Monte Carlo confidence intervals. *Biometrics* **40**, 811–817.

Buckland, S.T. (1987). On the variable circular plot method of estimating density. *Biometrika* **43**, 363–384.

Buker, J.B. and Groen, N.M. (1989). Distribution of Black-tailed Godwits *L. limosa* in different grassland types during the breeding season. *Limosa* **62**, 183–190.

Bullock, I.D. and Gomersall, C.H. (1981). The breeding populations of terns in Orkney and Shetland in 1980. *Bird Study* **28**, 187–200.

Bundy, G. (1978). Breeding Red-throated Divers in Shetland. *British Birds* **71**, 199–208.

Bunn, D.S., Warburton, A.B. and Wilson, R.D.S. (1982). *The Barn Owl*. T & AD Poyser, Calton.

Burnham, K.P. and Anderson, D.R. (1984). The need for distance data in transect counts. *Journal of Wildlife Management* **48**, 1248–1254.

Burnham, K.P., Anderson, D.R. and Laake, J.L. (1980). Estimation of density from line transect sampling of biological populations. *Wildlife Monographs* **72**, 1–200.

Cadbury, C.J. (1980). The status and habitats of the Corncrake in Britain 1978–79. *Bird Study* **27**, 203–218.

Cadbury, C.J. (1981). Nightjar census methods. *Bird Study* **28**, 1–4.

Campbell, L.H. and Talbot, T.R. (1987). Breeding status of Black-throated Divers in Scotland. *British Birds* **80**, 1–8.

Clark, N.A. (1990). *Distribution Studies of Waders and Shelduck in the Severn Estuary*. Report to UK Department of Energy's Renewable Energy Research and Development Programme (ETSU TID 4076), London.

Clarke, R. and Watson, D. (1990). The Hen Harrier *Circus cyaneus* winter roost survey in Britain and Ireland. *Bird Study* **37**, 84–100.

Clobert, J. Lebreton, J.D. and Allaine, D. (1987). A general approach to survival rate estimation by recaptures or resightings of marked birds. *Ardea* **75**, 133–142.

Cody, M.L. (1985). *Habitat Selection in Birds*. Academic Press, London.

Collar, N.J. and Andrew, P. (1988). *Birds to Watch: the ICBP World Checklist of Threatened Birds*. ICBP Technical Publication No. 8. ICBP, Cambridge.

Conroy, M.J., Hines, J.E. and Williams, B.K. (1989). Procedures for the analysis of bird-recovery data and user instructions for programme MULT. *Resource publication–US Fish and Wildlife Service* **175**, 1–61.

Cooch, E.G., Lark, D.B., Rockwell, R.F. and Cooke, F. (1989). Long term decline in fecundity in a Snow Goose population: evidence for density dependence. *Journal of Animal Ecology* **58**, 711–726.

Cormack, R.M. (1968). The statistics of capture–recapture methods. *Ocean Marine Biology Annual Review* **6**, 455–506.

Cormack, R.M. (1979). Models for capture–recapture, pp. 217–255. In: Cormack, R.M., Patil, G.P. and Robson, D.S. (eds.). *Sampling Biological Populations. Statistical Ecology Series. Vol 5.* International Co-op Publishing House, Fairland, Maryland, USA.

Cramp. S. and Simmons, K.E.L. (eds.) (1977 onwards). *The Birds of the Western Palearctic. Vols I–V.* Academic Press, London.

Day, J. (1988). Marsh Harriers in Britain. *RSPB Conservation Review* **2**, 17–19.

DeSante, D.F. (1981). Censusing technique in a California coastal scrub breeding bird community. *Studies in Avian Biology* **6**, 177–186.

DeSante, D.F. (1986). A field test of the variable circular-plot censusing method in a Sierran subalpine forest habitat. *Condor* **88**, 129–142.

del Nevo, A.J. (1990). *Reproductive and Feeding Ecology of Common Guillemots (Uria aalge) on Fair Isle, Shetland.* Ph.D. thesis. University of Sheffield.

Diamond, A.W., Gaston, A.J. and Brown, R.G.B. (1986). Converting PRIOP counts of seabirds at sea to absolute densities. *Progress Notes of Canadian Wildlife Service* **164**, 1–21.

Dixon, K.R. and Chapman, J.A. (1980). Harmonic mean measure of animal activity areas. *Ecology* **61**, 1040–1044.

Drennan, S.R. (1981). The Christmas Bird Count: an overlooked and underused sample. *Studies in Avian Biology* **6**, 24–29.

du Feu, C., Hounsome, M. and Spence, I. (1983). A single-session mark/recapture method of population estimation. *Ringing and Migration* **4**, 211–226.

Duebbert, H.F. and Lokemoen, J.T. (1976). Duck nesting in fields of undisturbed grass–legume cover. *Journal of Wildlife Management* **40**, 39–49.

Dunnet, G.M., Ollason, J.C. and Anderson, A. (1979). A 28-year study of breeding Fulmars *Fulmarus glacialis* in Orkney. *Ibis* **121**, 293–300.

Efron, B. (1982). *The Jackknife, the Bootstrap and Other Resampling Methods.* Society for Industrial and Applied Mathematics, Philadelphia.

Ellenberg, H. (1985). How to use species area relationships to compare grid-mapping results from different grid sizes, pp. 321–329. In: Taylor, K., Fuller, R.J. and Lack, P. (eds.). *Bird Census and Atlas Studies. Proceedings VIII International Conference on Bird Census and Atlas Work.* BTO, Tring, Herts.

Emlen, J.T. (1977). Estimating breeding season bird densities from transect counts. *Auk* **94**, 455–468.

Enemar, A. (1959). On the determination of the size and composition of a passerine bird population during the breeding season. *Vår Fågelvärld Supplement* **2**, 1–114.

Engstrom, R.T. and James, F.C. (1984). An evaluation of methods used in the Breeding Bird Census. *American Birds* **28**, 19–23.

Erdelen, M. (1984). Bird communities and vegetation structure I. Correlations and comparisons of simple diversity indices. *Oecologia* **61**, 277–284.

Evans, P.G.H. (1980). *Auk Censusing Manual.* British Seabird Group, Tring, Herts.

Evans, P.G.H. (1986). Monitoring seabirds in the North Atlantic. *NATO ASI Series* **G12**, 179–206.

Everett, M.J. (1982). Breeding Great and Arctic Skuas in Scotland in 1974–75. *Seabird Report* **6**, 50–58.

Ewins, P.J. (1985). Colony attendance and censusing of Black Guillemots *Cepphus grylle* in Shetland. *Bird Study* **32**, 176–185.

Fowler, J. and Cohen, L. (1986). *Statistics for Ornithologists.* BTO, Tring, Herts.

Fuller, R.J. (1982). *Bird Habitats in Britain.* T & AD Poyser, Calton.

Fuller, M.R. and Mosher, J.A. (1981). Methods of detecting and counting raptors. *Studies in Avian Biology* **6**, 235–246.

Fuller, R.J., Green, G.H. and Pienkowski, M.W. (1983). Field observations on methods used to count waders breeding at high density in the Outer Hebrides, Scotland. *Wader Study Group Bulletin* **39**, 27–29.

Fuller, R.J. and Langslow, D.R. (1984). Estimating numbers of birds by point counts: how long should counts last? *Bird Study* **31**, 195–202.

Fuller, R.J. and Marchant, J.H. (1985). Species-specific problems of cluster analysis in British mapping censuses, pp. 83–86. In: Taylor, K., Fuller, R.J. and Lack, P.C. (eds.). *Bird Census and Atlas Studies. Proceedings VIII International Conference on Bird Census and Atlas Work.* BTO, Tring, Herts.

Fuller, R.J., Marchant, J.H. and Morgan, R.A. (1985). How representative of agricultural practice in Britain are Common Bird Census farmland plots? *Bird Study* **32**, 56–70.

Fuller, R.J. and Moreton, B.D. (1987). Breeding bird populations of Kentish sweet chestnut (*Castanea sativa*) coppice in relation to age and structure of the coppice. *Journal of Applied Ecology* **24**, 13–27.

Fuller, R.J., Reed, T.M., Buxton, N.E., Webb, A., Williams, T.D. and Pienkowski, M.W. (1986). Populations of breeding waders Charadrii and their habitats on the crofting lands of the Outer Hebrides, Scotland. *Biological Conservation* **37**, 333–361.

Fuller, R.J., Stuttard, P. and Ray, C.M. (1989). The distribution of breeding songbirds within mixed coppiced woodland in Kent, England, in relation to vegetation age and structure. *Annales Zoologici Fennici* **26**, 265–275.

Furness, R.W. (1982). Methods used to census skua colonies. *Seabird Report* **6**, 11–17.

Gaston, A.J., Collins, B.T. and Diamond, A.W. (1987). Estimating densities of birds at sea and the proportion in flight from counts made on transects of indefinite width. *Canadian Wildlife Service Occasional Paper* **59**, 1–14.

Gauch, H.G. (1982). *The Use of Multivariate Analysis in Community Ecology.* Cambridge University Press, Cambridge.

Geissler, P.H. and Noon, B.R. (1981). Estimates of avian population trends from the North American Breeding Bird Survey. *Studies in Avian Biology* **6**, 45–61.

Goss-Custard, J.D. and Durell, S.E.A. le V. Dit (1990). Bird behaviour and environmental planning: approaches in the study of wader populations. *Ibis* **132**, 273–289.

Green, R.E. (1984). The feeding ecology and survival of partridge chicks (*Alectoris rufa* and *Perdix perdix*) on arable farmland in East Anglia. *Journal of Animal Ecology* **21**, 817–830.

Green, R.E. (1985a). Estimating the abundance of breeding Snipe. *Bird Study* **32**, 141–149.

Green, R.E. (1985b). *The Management of Lowland Wet Grasslands for Breeding Waders.* RSPB, Sandy, Beds.

Green, R.E. (1988). Effects of environmental factors on the timing and success of breeding of Common Snipe *Gallinago gallinago* (Aves: Scolopacidae). *Journal of Applied Ecology* **25**, 79–93.

Green, R.E. and Hirons, G.J.M. (1988). Effects of nest failure and spread of laying on counts of breeding birds. *Ornis Scandinavica* **19**, 76–78.

Green, R.E. and Hirons, G.J.M. (1990). The relevance of population studies to the conservation of threatened birds, pp. 595–631. In: Perrins, C.M., Lebreton, J.D. and Hirons, G.J.M. (eds.). *Bird Population Studies.* Oxford University Press, Oxford.

Gribble, F.C. (1983). Nightjars in Britain and Ireland in 1981. *Bird Study* **30**, 165–176.

Haila, Y. and Kuuesla, S. (1982). Efficiency of one-visit censuses of bird communities breeding on small islands. *Ornis Scandinavica* **13**, 17–24.

Hairston, N.G. (1989). *Ecological Experiments: Purpose, Design and Execution.* Cambridge University Press, Cambridge.

Hanssen, O.J. (1982). Evaluation of some methods for censusing larid populations. *Ornis Scandinavica* **13**, 183–188.

Harris, M.P. (1983). *The Puffin.* T & AD Poyser, Calton.

Harris, M.P. (1987). A low-input method of monitoring Kittiwake *Rissa tridactyla* breeding success. *Biological Conservation* **41**, 1–10.

Harris, M.P. (1988). Variation in the correction factor used for converting counts of individual Guillemots *Uria aalge* into breeding pairs. *Ibis* **131**, 85–93.

Harris, M.P. (1989). *Development of Monitoring of Seabird Populations and Performance.* Institute of Terrestrial Ecology: final report to Nature Conservancy Council, Peterborough.

Harris, M.P. and Forbes, R. (1987). The effect of date on counts of nests of Shags *Phalacrocorax aristotelis*. *Bird Study* **34**, 187–190.

Harris, M.P. and Lloyd, C.S. (1977). Variations in counts of seabirds from photographs. *British Birds* **70**, 200–205.

Harris, M.P. and Murray, S. (1981). Monitoring of Puffin numbers at Scottish colonies. *Bird Study* **28**, 15–20.

Harris, M.P. and Rothery, P. (1988). Monitoring of Puffin burrows on Dun, St Kilda, 1977–1987. *Bird Study* **35**, 97–99.

Hatch, S.A. and Hatch, M.A. (1989). Attendance patterns of Murres at breeding sites: implications for monitoring. *Journal of Wildlife Management* **53**, 483–493.

Hestbeck, J.B. and Malecki, R.A. (1989). Mark–resight estimate of Canada Goose midwinter numbers. *Journal of Wildlife Management* **53**, 749–752.

Heubeck, M., Richardson, M.G. and Dore, C.P. (1986). Monitoring numbers of Kittiwakes *Rissa tridactyla* in Shetland. *Seabird* **9**, 32–42.

Hildén, O. (1986). Long-term trends in the Finnish bird fauna: methods of study and some results. *Vår Fågelvärld Supplement* **11**, 61–69.

Hildén, O. (1987). Finnish winter bird censuses: long-term trends in 1956–84. *Acta Oecologica–Oecologica Generalis* **8**, 157–168.

Hill, D.A. (1982). *The Comparative Population Ecology of Mallard and Tufted Duck.* D.Phil. thesis, University of Oxford.

Hill, D.A. (1984a). Factors affecting nest success in the Mallard and Tufted duck. *Ornis Scandinavica* **15**, 115–122.

Hill, D.A. (1984b). Clutch predation in relation to nest density in Mallard and Tufted Duck. *Wildfowl* **35**, 151–156.

Hill, D.A. (1988). Population dynamics of the avocet (*Recurvirostra avosetta*) breeding in Britain. *Journal of Animal Ecology* **57**, 669–683.

Hill, D.A., Lambton, S.J., Proctor, I. and Bullock, I. (1991). Winter bird communities in woodland in The Forest of Dean, England, and some implications of livestock grazing. *Bird Study* **38**, 57–71.

Hill, D.A. and Robertson, P.A. (1988). *The Pheasant: Ecology, Management and Conservation.* Blackwell Scientific Publications, Oxford.

Hill, D.A., Taylor, S., Thaxton, R., Amphlet, A. and Horn, W. (1990). Breeding bird communities of native pine forest, Scotland. *Bird Study* **37**, 133–141.

Hirons, G.J.M. (1980). The significance of roding by Woodcock *Scolopax rusticola*: an alternative explanation based on observations of marked birds. *Ibis* **22**, 350–354.

Hirons, G.J.M. and Johnson, T.H. (1987). A quantitative analysis of habitat preferences of Woodcock *Scolopax rusticola* in the breeding season. *Ibis* **129**, 371–382.

Horne, J. and Short, J. (1988). A note on the sightability of Emus during an aerial survey. *Australian Wildlife Research* **15**, 647–649.

Housden, S., Thomas, G.T., Bibby, C.J. and Porter, R. (1991). Towards a habitat conservation strategy for bird habitats in Britain. *RSPB Conservation Review* **5**, 9–16.

Howes, J.R. (1987). *Rapid Assessment Techniques for Coastal Wetland Evaluation. Results of a Workshop held in Selangor, West Malaysia. 1–7 March 1987.* INTERWADER publication no. 24, Kuala Lumpur.

Hudson, P. (1986). *Red Grouse: The Biology and Management of a Wild Gamebird.* The Game Conservancy Trust, Fordingbridge.

Hudson, P. and Rands, M. (1988). *Ecology and Management of Gamebirds.* Blackwell Scientific Publications, Oxford.

Hughes, S.W.M., Bacon, P. and Flegg, J.J.M. (1979). The 1975 census of the Great Crested Grebe in Britain. *Bird Study* **26**, 213–226.

Iman, R.L. and Davenport, J.M. (1980). Approximations to the critical region of the Friedman statistic. *Community Statistics* **A9**, 571–595.

International Bird Census Committee (1969). Recommendations for an international standard for a mapping method in bird census work. *Bird Study* **16**, 248–255.

Ivlev, V.F. (1961). *Experimental Ecology of the Feeding of Fishes.* Yale University Press, New Haven, Connecticut.

Jacobs, J. (1974). Quantitative measurement of food selection. A modification of the forage ratio and Ivlev's electivity index. *Oecologia* **14**, 413–417.

James, F.C. and McCulloch, C.E. (1985). Data analysis and the design of experiments in ornithology, pp. 1–63. In: Johnston, R.F. (ed.). *Current Ornithology Vol. 2.* Plenum Press, New York.

James, F.C. and Shugart, H.H. (1970). A quantitative method of habitat description. *Audubon Field Notes* **24**, 727–736.

James, P.C. and Robertson, H.A. (1985). The use of playback recordings to detect and census nocturnal burrowing seabirds. *Seabird* **7**, 18–20.

Järvinen, O. and Väisänen, R.A. (1975). Estimating relative densities of breeding birds by the line transect method. *Oikos* **26**, 316–322.

Järvinen, O. and Väisänen, R.A. (1983a). Correction coefficients for line transect censuses of breeding birds. *Ornis Fennica* **60**, 97–104.

Järvinen, O. and Väisänen, R.A. (1983b). Confidence limits for estimates of population density in line transects. *Ornis Scandinavica* **14**, 129–134.

Johnson, D.H. (1980). The comparison of usage and availability measurements for evaluating resource preference. *Ecology* **61**, 65 71.

Jolly, G.M. (1965). Explicit estimates from capture–recapture data with both death and immigration-stochastic model. *Biometrika* **52**, 225–247.

Jouventin, P. and Weimerskirch, H. (1990). Satellite tracking of Wandering Albatrosses. *Nature* **343**, 746–748.

Kanyamibwa, S., Schierer, A., Pradel, R. and Lebreton, J.D. (1990). Changes in adult annual survival rates in a western European population of the White Stork *Ciconia ciconia. Ibis* **132**, 27–35.

Kendeigh, S.C. (1944). Measurement of bird populations. *Ecological Monographs* **14**, 67–106.

Kenward, R.E. (1987). *Wildlife Radio-tagging: Equipment, Field Techniques and Data Analysis.* Academic Press, London.

Kirby, J.S. (1987). *Birds of Estuaries Enquiry—Instructions to Counters.* BTO, Tring, Herts.

Kirby, J.S. (1990). *A Guide to Birds of Estuaries Enquiry Counting Procedure During the 1982/83 to 1988/89 Period, and Recommendations for the Future.* BTO, Tring, Herts.

Koskimies, P. and Väisänen, R.A. (1991). *Monitoring Bird Populations: a Manual of Methods applied in Finland.* Finnish Museum of Natural History, Helsinki, Finland.

Laake, J.L., Burnham, K.P. and Anderson, D.R. (1979). *User's Manual for Program TRANSECT.* Utah State University Press, Logan, Utah.

Lack, P. (1986). *The Atlas of Wintering Birds in Britain and Ireland.* T & AD Poyser, Calton.

Leslie, P.H. (1945). On the use of matrices in certain population mathematics. *Biometrika* **33**, 183–212.

Lincoln, F.C. (1930). Calculating waterfowl abundance on the basis of banding returns. *USDA Circular* **118**, 1–4.

Lindén, H. and Rajala, P. (1981). Fluctuations in long-term trends in the relative densities of tetraonid populations in Finland, 1964–1977. *Finnish Game Research* **39**, 13–24.

Lloyd, C., Tasker, M.L. and Partridge, K. (1991). *The Status of Seabirds in Britain and Ireland.* T & AD Poyser, Calton.

Lovvorn, J.R. (1989). Distributional responses of Canvasback Ducks to weather and habitat change. *Journal of Applied Ecology* **26**, 113–130.

MacArthur, R.H. and MacArthur, J.W. (1961). On bird species diversity. *Ecology* **42**, 594–598.

MacArthur, R.H., MacArthur, J.W. and Preer, J. (1962). On bird species diversity:

II. Prediction of bird census from habitat measurements. *American Naturalist* **96**, 167–174.

Marchant, J.H. (1983). *BTO Common Birds Census Instructions*. BTO, Tring, Herts.

Marchant, J.H., Hudson, R., Carter, S.P. and Whittington, P. (1990). *Population Trends in British Breeding Birds*. BTO, Tring, Herts.

Marquiss, M. (1989). Grey Herons *Ardea cinerea* breeding in Scotland: numbers, distribution, and census techniques. *Bird Study* **36**, 181–191.

Marquiss, M., Newton, I. and Ratcliffe, D.A. (1978). The decline of the Raven, *Corvus corax*, in relation to afforestation in southern Scotland and northern England. *Journal of Applied Ecology* **15**, 129–144.

Massa, R. and Fedrigo, A. (1989). A new approach for compiling a winter bird atlas by means of point counts. *Annales Zoologici Fennici* **26**, 207–212.

Matthysen, E. (1989). Nuthatch *Sitta europaea* demography, beech mast, territoriality. *Ornis Scandinavica* **20**, 278–282.

Mead, C. (1987). *Owls*. Whittet Books, London.

Meek, E.R., Booth, C.J., Reynolds, P. and Ribbands, B. (1983). Breeding skuas in Orkney. *Seabird* **7**, 21–29.

Miller, R.G. (1974). The Jackknife—a review. *Biometrika* **61**, 1–15.

Millsap, B.A. and LeFranc, M.N., Jr. (1988). Road transects for raptors: how reliable are they? *Journal of Raptor Research* **22**, 8–16.

Moss, R. and Oswald, J. (1985). Population dynamics of Capercaillie in a north-east Scottish glen. *Ornis Scandinavica* **16**, 229–238.

Mudge, G.P. (1988). An evaluation of current methodology for monitoring changes in the breeding populations of Guillemots *Uria aalge*. *Bird Study* **35**, 1–9.

Murray, S. and Wanless, S. (1986). The status of the Gannet in Scotland 1984–85. *Scottish Birds* **14**, 74–85.

NCC (1990). *Handbook for Phase 1 Habitat Survey: a Technique for Environmental Audit*. England Field Unit, Nature Conservancy Council, Peterborough.

Nettleship, D.N. (1976). Census techniques for seabirds of Arctic and Eastern Canada. *Canadian Wildlife Service Occasional Papers* **25**, 1–33.

Neu, C.W., Byers, C.R. and Peek, J.M. (1974). A technique for analysis of utilisation-availability data. *Journal of Wildlife Management* **38**, 541–545.

Newton, I. (1986). *The Sparrowhawk*. T & AD Poyser, Calton.

Newton, I. (1988). A key factor analysis of a Sparrowhawk population. *Oecologia* **76**, 588–596.

Newton, I., Wyllie, I. and Mearns, R. (1986). Spacing of Sparrowhawks in relation to food supply. *Journal of Animal Ecology* **55**, 361–370.

Nichols, J.D., Noon, B.R., Stokes, S.L. and Hines, J.E. (1981). Remarks on the use of mark–recapture methodology in estimating avian population size. *Studies in Avian Biology* **6**, 121–136.

O'Connor, R.J. and Mead, C.J. (1984). The Stock Dove in Britain, 1930–80. *British Birds* **77**, 181–201.

O'Connor, R.J. and Shrubb, M. (1986). *Farming and Birds*. Cambridge University Press, Cambridge.

Ogilvie, M.A. (1986). The Mute Swan *Cygnus olor* in Britain 1983. *Bird Study* **33**, 121–137.

Ormerod, S.J., Tyler, S.J., Pester, S.J. and Cross, A.V. (1988). Censusing distribution and population of birds along upland rivers using measured ringing effort: a preliminary study. *Ringing and Migration* **9**, 71–82.

Osborne, P.E. and Tigar, B. (1992). Interpreting bird atlas data using logistic models: an example from Lesotho, South Africa. *Journal of Applied Ecology* **29**, 55–62.

Österlöf, S. and Stolt, B.-O. (1982). Population trends indicated by birds ringed in Sweden. *Ornis Scandinavica* **13**, 135–140.

Otis, D.L., Burnham, K.P., White, G.C. and Anderson, D.R. (1978). Statistical inference from capture data on closed animal populations. *Wildlife Monographs* **62**, 1–135.

Owen, M. (1971). The selection of feeding sites by White-fronted Geese in winter. *Journal of Applied Ecology* **8**, 905–917.

Owen, M., Atkinson-Willes, G.L. and Salmon, D.G. (1980). *Wildfowl in Great Britain.* Cambridge University Press, Cambridge.

Owen, M. and Black, J.M. (1989). Factors affecting the survival of Barnacle Geese on migration from the breeding grounds. *Journal of Animal Ecology* **58**, 603–617.

Palmer, R.S. (ed.) (1962 onwards). *Handbook of North American Birds. Vols. I–V.* Yale University Press, New Haven and London.

Parrinder, E.D. (1989). Little Ringed Plovers *Charadrius dubius* in Britain in 1984. *Bird Study* **36**, 147–153.

Peach, W. and Baillie, S.R. (1989). Population changes on constant effort sites, 1987–1988. *BTO News* **161**, 12–13.

Petty, S.J. and Avery, M.I. (1990). Forest Bird Communities. *Forestry Commission, Occasional Paper* **26**, 1–110.

Pollock, K.H. (1981). Capture–recapture models: a review of current methods, assumptions and experimental design. *Studies in Avian Biology* **6**, 426–435.

Pomeroy, D. (1989). Using East African bird atlas data for ecological studies. *Annales Zoologici Fennici* **26**, 309–314.

Potts, G.R. (1986). *The Partridge: Pesticides, Predation and Conservation.* Collins, London.

Pöysä, H. (1984). Temporal and spatial dynamics of waterfowl populations in a wetland area—a community ecological approach. *Ornis Fennica* **61**, 99–108.

Prater, A.J. (1979). Trends in accuracy of counting birds. *Bird Study* **26**, 198–200.

Prater, A.J. (1981). *Estuary Birds of Britain and Ireland.* T & AD Poyser, Calton.

Prater, A.J. (1989). Ringed plover *Charadrius hiaticula* breeding population of the United Kingdom in 1984. *Bird Study* **36**, 154–161.

Quade, D. (1979). Using weighted rankings in the analysis of complete blocks with additive block effects. *Journal of American Statistics Association* **74**, 680–683.

Ralph, C.J. and Scott, J.M. (eds.) (1981). *Estimating the Number of Terrestrial Birds: Studies in Avian Biology no. 6.* Cooper Ornithological Society, Lawrence, Kansas, USA.

Rapold, C., Kersten, M. and Smit, C. (1985). Errors in large scale shorebird counts. *Ardea* **73**, 13–24.

Reed, T.M. and Fuller, R.J. (1983). Methods used to assess populations of breeding waders on machair in the Outer Hebrides. *Wader Study Group Bulletin* **39**, 14–16.

Reed, T.M., Barrett, J.C., Barrett, C. and Langslow, D.R. (1984). Diurnal variability in the detection of dunlin *Calidris alpina. Bird Study* **31**, 245–246.

Reed, T.M., Barrett, C., Barrett, J., Hayhow, S. and Minshull, B. (1985). Diurnal variability in the detection of waders on their breeding grounds. *Bird Study* **32**, 71–74.

Reynolds, C.M. (1979). The heronries census: 1972–1977 population changes and a review. *Bird Study* **26**, 7–12.

Reynolds, R.T., Scott, J.M. and Nussbaum, R.A. (1980). A variable circular plot method for estimating bird numbers. *Condor* **82**, 309–313.

Richardson, M.G. (1990). The distribution and status of Whimbrel *Numenius p. phaeopus* in Shetland and Britain. *Bird Study* **37**, 61–68.

Robbins, C.S. (1981). Effect of time of day on bird activity. *Studies in Avian Biology* **6**, 275–286.

Robbins, C.S., Bystrak, D. and Geissler, P.H. (1986). The Breeding Bird Survey: its first fifteen years, 1965–1979. *United States Department of the Interior, Fish and Wildlife Service, resource publication* **157**, 1–196.

Robbins, C.S., Droege, S. and Sayer, J.R. (1989). Monitoring bird populations with Breeding Bird Survey and atlas data. *Annales Zoologici Fennici* **26**, 279–304.

Robertson, P.A., Woodburn, M.I.A., Bealey, C.E., Ludolf, I.C. and Hill, D.A. (1990). *Pheasants and Woodlands: Habitat Selection, Management and Conservation.* Report to the Forestry Commission. The Game Conservancy, Fordingbridge.

Rodwell, J.S. (ed.) (1991). *British Plant Communities: vol. 1. Woodlands and Scrub.* Cambridge University Press, Cambridge.

Rolstad, J. and Wegge, P. (1987). Distribution and size of Capercaillie leks in relation to old forest fragmentation. *Oecologia* **72**, 389–394.

Root, T. (1988). *Atlas of Wintering North American Birds. An Analysis of Christmas Bird Count Data.* University of Chicago Press, Chicago.

Rose, P. (1990). *Manual for International Waterfowl Census Coordinators.* International Waterfowl and Wetlands Research Bureau, Slimbridge.

Rotenberry, J. (1985). The role of habitat in avian community composition: physioguomy or floristics? *Oecologia* **67**, 213–217.

Rotenberry, J.T. and Wiens, J.A. (1980). Habitat structure, patchiness, and avian communities in North American steppe vegetation: a multivariate analysis. *Ecology* **61**, 1228–1250.

Rotenberry, J.T. and Wiens, J.A. (1985). Statistical power analysis and community wide patterns. *The American Naturalist* **125**, 164–168.

Rumble, M.A. and Flake, L.D. (1982). A comparison of two waterfowl brood survey techniques. *Journal of Wildlife Management* **46**, 1048–1053.

Ryan, P.G. and Cooper, J. (1989). The distribution and abundance of aerial seabirds in relation to Antarctic krill in the Pryds Bay region, Antarctica, during late Summer. *Polar Biology* **10**, 199–209.

Sage, B.L. and Vernon, J.D.R. (1978). The 1975 National survey of Rookeries. *Bird Study* **25**, 64–86.

Salmon, D.G. (1989). In: Prys-Jones, R. and Kirby, J. (eds.). *Wildfowl and Wader Counts 1988–89.* Wildfowl and Wetlands Trust, Slimbridge.

Schwaller, M.R., Olser, C.E., Zhenqui Ma, Zhiliang Zhu and Dahmer, P. (1989). A remote sensing analysis of Adelie penguin rookeries. *Remote Sensing of Environment* **28**, 199–206.

Scott, M.J. and Ramsay, F.L. (1981). Length of count period as possible source of bias in estimating bird densities. *Studies in Avian Biology* **6**, 409–413.

Scott, M.J., Ramsay, F.L. and Kepler, C.B. (1981). Distance estimation as a variable in estimating bird numbers. *Studies in Avian Biology* **6**, 334–341.

Scott, P. (1981). *Variation of Bill— Markings of Migrant Swans Wintering in Britain.* The Wildfowl Trust, Slimbridge.

Seabird Group/NCC (1988). *Seabird Colony Register: Recommended Methods for Counting Breeding Seabirds.* Seabird Group/NCC, Peterborough, U.K.

Seber, G.A.F. (1965). A note on the multiple-recapture census. *Biometrika* **52**, 249.

Seber, G.A.F. (1973). *The Estimation of Animal Abundance.* Hafner, New York and Griffin, London.

Sharrock, J.T.R. (1976). *The Atlas of Breeding Birds in Britain and Ireland.* T & AD Poyser, Calton.

Shawyer, C.R. (1987). *The Barn Owl in Britain: Its Past, Present and Future.* The Hawk Trust, London.

Shrubb, M. and Lack, P.C. (1991). The numbers and distribution of Lapwings *V. vanellus* nesting in England and Wales in 1987. *Bird Study* **38**, 20–38.

Sitters, H. (ed.). (1988). *Tetrad Atlas of the Breeding Birds of Devon.* Devon Birdwatching and Preservation Society, Yelverton.

Smith, J.M. (1988). Landsat TM study of afforestation in northern Scotland and its impact on breeding bird populations, pp. 1369–1370. In: Gvyenne, T.D. and Hunt, J.J (eds.). *Remote sensing. Proceedings IGARSS '88 symposium, Edinburgh. Vol. 3.* European Space Agency, ESTEC, Noordwijk, ESA, SP-284.

Smith, K.W. (1983). The status and distribution of waders breeding on lowland wet grasslands in England and Wales. *Bird Study* **30**, 177–192.

Southern, H.N. and Lowe, V.P.W. (1968). Pattern of distribution of prey and predation in Tawny Owls. *Journal of Animal Ecology* **37**, 75–97.

Southwood, T.R.E. (1978). *Ecological Methods.* 2nd Edn. Chapman and Hall, London.

SOVON (1987). *Atlas van de Nederlandse Vogels.* SOVON, Arnhem.

Stowe, T.J. (1982). Recent population trends in cliff-breeding seabirds in Britain and Ireland. *Ibis* **124**, 502–510.

Stowe, T.J. and Hudson, A.V. (1988). Corncrake studies in the western isles. *RSPB Conservation Review* **2**, 38–42.

Tappe, P.A., Whiting, R.M. and George, R.R. (1989). Singing-ground surveys for Woodcock in East Texas. *Wildlife Society Bulletin* **17**, 36–40.

Tapper, S. (1989). The 1989/90 shooting season. Pp. 28–34. In: *The Game Conservancy Review of 1989*. Nodder, C. (ed.). Game Conservancy, Fordingbridge.

Tasker, M.L., Hope Jones, P., Dixon, T. and Blake, B.F. (1984). Counting seabirds at sea from ships: a review of methods employed and a suggestion for a standardized approach. *Auk* **101**, 567–577.

Taylor, K., Hudson, R. and Horne, G. (1988). Buzzard breeding distribution and abundance in Britain and Ireland in 1983. *Bird Study* **35**, 109–118.

Thompson, J.J. (1989). A comparison of some avian census techniques in a population of Lovebirds at Lake Naivasha, Kenya. *African Journal of Ecology* **27**, 157–166.

Thompson, K.R. and Rothery, P. (1991). A census of Black-browed Albatross *Diomedea melanophyrs* population on Steeple Jason Island, Falkland Islands. *Biological Conservation* **56**, 39–48.

Udvardy, M.D.F. (1981). An overview of grid-based atlas works in ornithology. *Studies in Avian Biology* **6**, 103–109.

Underhill, L.G. and Fraser, M.W. (1989). Bayesian estimate of the number of Malachite Sunbirds feeding at an isolated and transient nectar resource. *Journal of Field Ornithology* **60**, 382–387.

Verner, J. (1985). Assessment of counting techniques, pp. 247–301. In: Johnston, R.F. (ed.). *Current Ornithology Vol. 2*. Plenum Press, New York.

Verner, J. and Milne, K.A. (1989). Coping with sources of variability when monitoring population trends. *Annales Zoologici Fennici* **26**, 191–199.

Vinicombe, K. (1982). Breeding and population of the Little Grebe. *British Birds* **75**, 204–218.

Wanless, S. and Harris, M.P. (1984). Effects of date on counts of nests of Herring and Lesser Black-backed Gulls. *Ornis Scandinavica* **15**, 89–94.

Watson, A., Payne, S. and Rae, R. (1989). Golden Eagles *Aquila chrysaetos*: land use and food in north east Scotland. *Ibis* **131**, 336–348.

Welsh, D.A. (1989). A report on breeding bird atlases in Canada. *Annales Zoologici Fennici* **26**, 305–308.

Whilde, A. (1985). The 1984 all-Ireland tern survey. *Irish Birds* **3**, 1–32.

Whittaker, R.H. (1977). Evolution of species diversity in land communities. *Evolutionary Biology* **10**, 1–67.

Wiens, J.A. (1969). An approach to the study of ecological relationships among grassland birds. *Ornithological Monographs* **8**, 1–93.

Wiens, J.A. (1973). Pattern and process in grassland bird communities. *Ecological Monographs* **43**, 237–270.

Wiens, J.A. (1981). Scale problems in avian censusing. *Studies in Avian Biology* **6**, 513–521.

Wiens, J.A. (1985). Habitat selection in variable environments: shrubsteppe birds, pp. 227–251. In: Cody, M.L. (ed.). *Habitat Selection in Birds*. Academic Press, New York.

Wiens, J.A. (1989). *The Ecology of Bird Communities, vols. 1 and 2*. Cambridge University Press, Cambridge.

Wiens, J.A. and Rotenberry, J.T. (1985). Response of breeding passerine birds to rangeland alteration in a North American shrubsteppe locality. *Journal of Applied Ecology* **22**, 655–668.

William, A.B. (1936). The composition and dynamics of a beech–maple climax community. *Ecological Monographs* **6**, 317–408.

Williamson, K. (1964). Bird census work in woodland. *Bird Study* **11**, 1–22.

Williamson, K. (1968). Buntings on a barley farm. The bird community of farmland. *Bird Study* **15**, 34–37.

Wilson, H.J. (1982). Movements, home ranges and habitat use of wintering Wood-cock in Ireland, pp. 168–178. In: Dwyer, T.J. and Storm, G.L. (eds.). *Papers of the Seventh Woodcock Symposium. Wildlife Research Report no. 14.* United States Department of the Interior: Fish and Wildlife Service, Washington D.C.

Woolhead, J. (1987). A method for estimating the number of breeding pairs of Great Crested Grebes *Podiceps cristatus* on lakes. *Bird Study* **34**, 82–86.

Wormell, P. (1976). The Manx Shearwaters of Rhum. *Scottish Birds* **9**, 103–118.

Yalden, D.W. and Yalden, P.E. (1989). The sensitivity of breeding Golden Plovers *Pluvialis apricaria* to human intruders. *Bird Study* **36**, 49–55.

Young, A.D. (1989). Spacing behaviour of visual and tactile-feeding shorebirds in mixed species groups. *Canadian Journal of Zoology* **67**, 2026–2028.

Species Index

General Index